「学者文库」

基于新型可充电锂电池固态和液态电解质的研究

梅新艺 著

吉林大学出版社

图书在版编目（CIP）数据

基于新型可充电锂电池固态和液态电解质的研究 / 梅新艺著 . —长春：吉林大学出版社，2020.2
ISBN 978－7－5692－6138－7

Ⅰ.①基… Ⅱ.①梅… Ⅲ.①锂电池—蓄电池—固体电解质电池—研究②锂电池—胶体蓄电池—研究 Ⅳ. ①TM911

中国版本图书馆 CIP 数据核字（2020）第 028848 号

书　　名	基于新型可充电锂电池固态和液态电解质的研究 JIYU XINXING KECHONGDIAN LIDIANCHI GUTAI HE YETAI DIANJIEZHI DE YANJIU
作　　者	梅新艺　著
策划编辑	李潇潇
责任编辑	李潇潇
责任校对	刘守秀
装帧设计	中联华文
出版发行	吉林大学出版社
社　　址	长春市人民大街 4059 号
邮政编码	130021
发行电话	0431－89580028/29/21
网　　址	http：//www.jlup.com.cn
电子邮箱	jdcbs@jlu.edu.cn
印　　刷	三河市华东印刷有限公司
开　　本	710mm×1000mm　1/16
印　　张	13.5
字　　数	210 千字
版　　次	2020 年 2 月第 1 版
印　　次	2020 年 2 月第 1 次
书　　号	ISBN 978－7－5692－6138－7
定　　价	95.00 元

版权所有　　翻印必究

Acknowledgement

First of all, I would like to express my deepest gratitude to my advisor Prof. Braja K. Mandal, for his support, patience, encouragement, inspiration and guidance throughout my Ph. D. study. It was a great honor to work under his supervision.

My sincere appreciation also goes to my thesis committee members: Prof. Carlo Segre, Prof. David Minh, and Prof. Hyun – Soon Chong for their time, helpful comments, and advice.

I am thankful to the Chemistry Department for providing me financial support throughout my program. Special thanks to my academic colleagues in my group, for giving me unforgettable memories.

To my parents, without your unconditional support and encouragement, this journey on the track of my graduate research career would have been impossible. Thank you very much.

I am dedicating this thesis to my grandfather and share the joy of this accomplishment. He has always been concerning about me during my adolescence.

Lastly, I would like to give my regards to my friends who supported me in many ways during the completion of the research work.

简　介

目前，可充电锂电池正广泛应用于我们的日常消费电子产品中，包括手机、笔记本电脑、相机等。锂电池具有极高的能量和功率密度，在电动汽车和混合动力汽车上也有着巨大的应用潜力。然而，目前锂电池在商业化应用中主要面临的问题有：较高的成本、溶剂易燃易爆的安全问题以及在较低温度下导电性能较差等，这些问题都有待寻找更好的解决方案。

围绕锂电池出现的上述问题，这篇博士论文展开了系统性的研究。本论文中主要研究讨论了两种可充电锂电池：锂离子电池和锂硫电池。本论文提出了两种有效的研究方案，可以进一步加强和优化可充电锂电池中电解质体系的实用性能。

第一种研究方案是基于传统的聚环氧乙烷（PEO）的固体聚合物电解质（SPE）体系。该方案主要研究了纳米颗粒级锂盐（NPLS）和低晶格能氟化二锂盐的制备。由于该方案中设计的固体聚合物电解质体系的玻璃化转变温度（T_g）明显降低，使得基于该固体聚合物电解质体系的离子电导率显著提高。

锂硫电池在工作过程中，阴极的多硫化物中间体极易溶解到电池的电解液中，从而产生多硫化物穿梭现象（PSS）。如何更好地遏制这种多硫化物穿梭现象，一直是锂硫电池研究领域的痛点。

本论文在第二种研究方案中精心设计了多个系列的室温离子液体

(RTILs)，用于有效遏制多硫化物穿梭现象。该方案中首先研究设计了三组双阳离子离子液体(DILs)，并完成了相关的合成和表征。

基于双阳离子离子液体的电解质体系具有良好的阻燃性、电化学稳定性和热稳定性。接着该方案中研究设计了12种新型非对称氟化室温离子液体（FRTILs），也完成了相关的合成和表征。

最后研究比对表明：基于新型非对称氟化室温离子液体的电解质体系的综合性能要优于基于双阳离子离子液体的电解质体系。

目 录
CONTENTS

LIST OF SYMBOLS ··· 1

ABSTRACT ··· 2

Chapter 1　Introduction ·· 4

　1.1　Introduction to lithium-ion batteries (LIB) ··················· 4

　1.2　Introduction to lithium-sulfur batteries (LSB) ················ 16

　1.3　Introduction to ionic liquids ······································ 20

　1.4　Introduction to measurements ·································· 26

　1.5　New Electrolytes in this thesis ·································· 28

Chapter 2　Active Nano-Particles Filled SPEs ··············· 30

　2.1　Introduction ··· 30

　2.2　Experimental ··· 36

　2.3　Results and Discussion ·· 41

　2.4　Summary ·· 59

Chapter 3　New Plasticized Low Lattice Energy Lithium Salts ······ 60

　3.1　Introduction ··· 60

　3.2　Experimental ··· 61

3.3	Results and Discussion	66
3.4	Summary	71

Chapter 4 New Nonflammable Di-cationic Ionic Liquids ············ 73

4.1	Introduction	73
4.2	Experimental	78
4.3	Results and Discussion	99
4.4	Summary	116

Chapter 5 New Asymmetric Fluorinated Room Temperature Ionic Liquids ············ 117

5.1	Introduction	117
5.2	Experimental	120
5.3	Results and Discussion	132
5.4	Summary	144

Chapter 6 Conclusion ············ 146

APPENDIX FTIR AND NMR SPECTRA ············ 148

BIBLIOGRAPHY ············ 188

LIST OF SYMBOLS

Symbol	Definition
σ	Ionic conductivity
δ	Chemical shift
$CDCl_3$	Deuterated chloroform
cm^{-1}	Wave number
°C	Degree in celsius
d	Doublet
DCM	Dimethanechloride
^1H	Proton
m	Multiplet
MHz	Mega Hertz, 10^6 Hertz
mg	Milligram, 10^{-3} grams
ml	Milliliter, 10^{-3} liters
M	Mole per liter
mV	Millivolts, 10^{-3} Volts
mS	MilliSiemens
S	Siemens
t	Triplet
THF	Tetrahydrofuran
V	Volts

ABSTRACT

Currently, rechargeable lithium batteries are widely used in our consumer electronic products, including cell phones, laptop computers, and cameras and so on. They have extraordinary potential for application in electric and hybrid electric vehicles due to their high energy and power density [1]; however, the major challenges including the higher cost, safety issues related to the solvents and conductibility at lower temperatures are still waiting to be fixed.

In this Ph. D. thesis, two types of rechargeable lithium batteries: lithium-ion batteries and lithium-sulfur batteries are discussed. Two different approaches are presented, in the direction of achieving an enhanced electrolyte system for rechargeable lithium batteries.

One approach is based on the conventional poly (ethylene oxide) (PEO) -based solid polymer electrolyte (SPE) system. The key feature of this approach is the preparation of nanoparticle lithium salts (NPLS) and low lattice energy fluorinated di-lithium salts. The ionic conductivities of these PEO-based SPEs were markedly improved, due to a decrease in the glass transition temperature (T_g) of the polymer.

For lithium-sulfur (Li-S) batteries, the polysulfide shuttle (PSS), caused by the dissolution of cathode polysulfide intermediates into the electrolyte, has delivered a mortal blow to nearly every attempt at obtaining a viable Li-S battery. So, another approach involves the strategic design and synthesis of a series of room temperature ionic liquids (RTILs) to prevent PSS: i) Three series of di-cationic ionic liquids (DILs) were synthesiyed and characterized. DILs-based electrolytes display excellent properties, such as non-flammability, high electrochemical stability and thermal stability. ii) Twelve new asymmetric fluorinated RTILs (FRTILs) were also introduced. The FRTILs based electrolytes show even better properties than DILs-based electrolytes.

Chapter 1   Introduction

1.1   Introduction to lithium-ion batteries (LIB)

Nowadays, people are facing severe environmental problems, such as running out of fossil fuels, fluctuation of oil prices, global warming, and pollution impact. With growing demanding of energy, it becomes more and more necessary to find alternative energy storage systems (ESS) [2].

To utilize other sustainable energy sources, for instance, solar and the wind, it is necessary to develop effective ESS to ensure the energy delivery and mediate power fluctuations [3]. Rechargeable lithium ion batteries (LIBs) have been considered as one of the most promising ESS, which are widely used in mobile electronic equipment, such as laptop computers, cellular phones, and digital cameras and so on [4], because of their high specific energy density (up to 150 Wh/kg), long cycle life and, highefficiency advantages.

Lithium-ion batteries have been manufactured yearly, based mainly on the use of $LiPF_6$ dissolved in liquid organic aprotic solvents, a major drawback for commercial production problem of leakage which still compromises safety and forces a non-flexible design. Rechargeable solid lithium batteries (RSLB) are particularly attractive because they combine high power density from the Li/Li$^+$ electrochemical couple with the safety and durability characteristics of plastics. The currently best available salt [5], Lithium bis (trifluoromethyl sulfonyl) imide (LiTFSI) is used with high molecular weight polyethylene oxide (PEO; MW: 4,000,000), but the SPE functions poorly below 60 ℃, so the salt requires the added plasticizers to maintain the amorphous character of the polymer host. As for LiTFSI, which is more expensive than LiPF$_6$, the main drawback to its use in lithium-ion batteries is related to corrosion of the aluminum current collector of the positive electrode. Another drawback in using lithium polymer batteries is their low cationic transference numbers in poly (oxyethylene) based host polymers [6]. The high anion mobility induces a concentration gradient for both species through the electrolyte, which results in increased internal resistance and voltage losses [7]. LiTFSI does not provide adequate electrolyte characteristics, viz., (i) cationic conductivity of at least 10^{-2} mS/cm at room temperature, (ii) wide electrochemical stability (0 – 5 V), (iii) safety through solvent-free design and lower reactivity with the lithium anode [8].

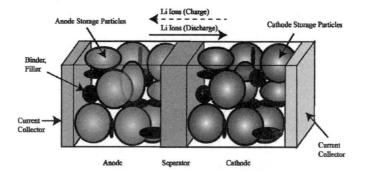

Figure 1.1. A schematic of the structure of the lithium ion battery [9].

In the lithium ion battery, there are positive and negative electrodes, called cathode and the anode respectively. The cathode is the positive electrode, is made of pure lithium metal oxide. The more uniform the composition, the better the performance and the longer battery life. The anode, a negative electrode which is located on the other side, is made of graphite, full of carbon with layer structure. The battery is filled with transport medium: the electrolyte. So the lithium ion carrying lithium charge can flow freely. The electrolyte may extremely pure, as free of water as possible for the purpose of assuring efficient charging and discharging. Between the two electrodes, there is a layer, the separator, which could prevent the short circuit. To the tiny lithium ions, the separator is permittable. The positively charged lithium ions pass from the cathode through the separator into the layered graphite structure of the anode, where they stored now the battery is charged. When the battery discharges that is the energy removed from the cell, the lithium ions travel via the electrolyte to the anode through the separator back to the cathode, so the light is lighted up.

Until now continually researches are on improving battery performance, cycle life, safety, lightness, shape design variety, no leakage system, and low volatility. Also to meet the requirement for applications in electric vehicles (EV) and hybrid electric vehicles (HEV), it is necessary to develop high energy density and low-cost materials.

For decades, many polymers, for instance, poly (acrylonitrile), poly (methyl methacrylate), poly (vinylidene fluoride-co-hexafluoropropylene), and poly (vinylidene fluoride) (PVdF) have been used as mediums for polymer electrolytes by soaking or swelling polymeric matrix with a liquid electrolyte. Nevertheless, the limit of this process lies in the poor stability and low electrolyte retention capability [10]. Since liquid organic solvents in the electrolytes significantly decrease lithium battery lifetime and safety. Alternatively, solid polymer electrolytes (SPEs) with promising electrochemical properties (e. g., conductivity, interfacial stability, and ionic transport properties) [11] and mechanical properties (e. g., viscous and elastic moduli, yield stress) provide an ideal way to solve the safety issue in the secondary lithium battery application [12].

The batteries based on metallic lithium anode and solid polymer electrolytes have encountered several problems: (i) A low ambient conductivity. (ii) Side reaction between electrolyte and lithium metal, giving rise to capacity loss during cycling. (iii) Safety hazards caused by dendrite formation during cycling. The conventional poly (ethylene oxide) (PEO) -based solid polymer electrolytes have been studied comprehensively due to the ability to form complexes with a wide variety of lithium salts [13].

1.1.1 Introduction to electrolytes

Electrolytes are a major concern related to lithiumion batteries for the reason that their properties, such as viscosity, ionic conductivity, and thermal stability, are capable of determining the performance of a battery. Polymer electrolytes now are expected to be the alternative electrolytes to satisfy the needs for power storage device as well as electric vehicles with high efficiency, high energy density, and long life. Polymer electrolytes can be divided into two types: liquid/ gel electrolytes and solid polymer electrolytes.

Gel electrolytes formed by immersing a significant amount of organic liquid electrolytes into polymer framework have very high ambient temperature ionic conductivities, but they still suffer from several disadvantages, such as worsening of mechanical properties, increased reactivity towards the metal electrode and release of volatiles.

On the other hand, the solid polymer electrolytes have commonly been prepared by dissolving the salt into the polymer framework as the case in the conversant polyethylene oxide (PEO) membrane. Finding the ideal SPEs maintaining good transport properties give the impression to be a challenging task, most solid electrolyte systems could not perform as well as liquids at room temperature, owing to their lower segmental chain mobility in the crystalline regions compared to the amorphous regions, even though various modification methods could be adopted to decrease the crystallinity of the polymer, e. g. , by adding inorganic fillers, forming a copolymer [14].

1.1.2 Introduction to liquid/gel electrolytes

Currently, organic solvents based electrolytes are most commonly used in LIB, because of their excellent properties such as high conductivity, low viscosity, and high ionic mobility. Aurbach et al. reported that when selecting electrolytes and their composition, the conductivity, electrochemical stability, operating temperature range, and safety concerns should be well considered. Satisfying all of these demands by using a single solvent is nearly impossible. Therefore, solvents exhibiting different physicochemical properties are frequently mixed for use in a variety of lithiumion battery applications [15].

At present, most commercial electrolytes for LIBs are composed of lithium salts, such as $LiPF_6$, and mixed solvents ethylene carbonate (EC) and linear carbonates, such as dimethyl, diethyl and ethyl-methyl carbonates (DMC, DEC, EMC, respectively) [16]. Cyclic EC with a high dielectric constant is a key component for the lithium salt dissociation and sufficient negative electrode passivation. However, EC offers a poor low-temperature performance. To extend the usable range of the electrolyte, linear carbonates essential to be added are, however, highly flammable because of their very low flash point and high vapor pressure [17].

Propylene carbonate (PC) is also a promising alternative to the state of theart electrolyte solvents attributable to its excellent properties like low melting point (55 ℃), high boiling point (240 ℃), and high flash point (132 ℃). Additionally, it enables high conductivities and complete salt dissociation due to its high permittivity (64.92 at 25 ℃), which is close to that of EC (89.78 at 25 ℃) as well as its moderate viscosity (2.53 mPa s at 25 ℃) [18]. Unfortunately, PC is not capable of forming a good SEI on graphite electrodes [19].

The resulting binary or mixed ternary solutions can achieve more rapid ion transport as a result of lower viscosity and subsequently higher ionic conductivity. The choice of co-solvent is a critical issue that significantly affects not only the conductivity but also the electrochemical performance, including the physicochemical properties of solid electrolyte interface (SEI) formation [20].

Liquid/gel electrolyte batteries show high ionic conductivity (10^{-2} ~ 10^{-3} S/cm). Nonetheless, it limits temperature range, low energy density, lack of mechanical stability, corrosion of the electrodes and growth of lithium dendrites from the anode to the cathode. Using liquid electrolytes also may perhaps lead to safety hazards such as fire and explosion, which can be caused by leakage of a fluid electrolyte comprising highly flammable linear carbonates as a solvent [21].

1.1.3 Introduction to solid polymer electrolytes

Over the past 30 years, solid polymer electrolytes (SPEs) have been intensively studied for the applications in solid-state lithium rechargeable batteries, which are predominantly motivated by their superior advantages over liquid electrolytes, such as avoiding high volatility, fluid leakage, and flammability of traditional liquid electrolytes or gel polymer. They also possess advantageous features of light weight, shape versatility, and easy processability. Therefore, they are promising components for high energy density power sources [13].

Conventional SPEs are typically prepared by dissolving lithium salt in a polymer matrix, in some cases, additionally containing plasticizers. The polymer matrix need contain a unit of Lewis base, usually ethylene oxide unit ($-OCH_2CH_2-$), to solvate lithium salt. It is generally acknowledged that the motion of Li ions is coupled to the motion of the polymer backbone and/or segment through the coordinating interactions between mobile Li^+ cations and Lewis base (e. g., ether oxygen atoms) [22].

Several polymer materials such as poly (vinylidene fluoride) (PVdF), polyacrylonitrile (PAN), poly (methyl methacrylate) (PMMA), polyurethane (PU), and polyvinyl chloride (PVC), have been used as the host polymers for the preparation of polymer electrolytes. However, one major limitation to the SPEs applications is their electrochemical and mechanical properties. To obtain a balance for the compatibility of ionic conductivity and mechanical properties, numerous investigations have been carried out such as the synthesis of the new polymer matrix, the preparation of polymer single-ion conductor, and the doping with nanoparticles. Then through blocking, grafting, crosslinking, compositing and blending methods [23], researchers can improve the performance of the polymer electrolyte by reducing the crystallinity of the polymer, increasing the concentration of ions and the proportion of the amorphous region contained in the system, along with decreasing glass transition temperature (T_g) of the polymer electrolyte system and improving the capacity of lithium ion dissociation. SPEs, which consist of polyethylene oxide (PEO) in which a lithium salt is dissolved, would be an ideal candidate [21].

1.1.4 Introduction to PEO

The development of solid polymer electrolytes began in the early 1970's with the discovery of complex ionic formation of PEO with alkali metal salts by P. V. Wright [24] [25]. Since then, the numbers of researches to PEO based SPEs as promising candidates to prepare thinner, lighter and safer LIBs have grown enormously [26]. Owing to the excellent chain flexibility and polar groups; therefore, the polarity of PEO causes lithium salts to be easily dissolved and promotes the smooth movement of dissociated ions.

Figure 1.2 Schematic representation of lithium ion migration associated with the segmental mobility of the PEO chains.

As is known, a Li-ion is coordinated with about four ether oxygen atoms of the same or different PEO chains. Lithium cation mobility in the solid phase is regulated by lithium cation-polymer interaction involving lithium cation with oxygen, nitrogen or sulfur coordination bonding, and can transport the dissociated ions by the segmental motion of the main chain. However, the lithium ion transportation can only happen in the amorphous phase of polymer hosts. Usually, the crystalline phase has much lower ionic conductivity than the amorphous phase [27], due to the segmental mobility of the polymer matrix that is frozen at the crystallinity phase. PEO based electrolytes exhibit destitute ambient temperature ionic conductivity due to the very high degree of crystallinity present in PEO at room temperature and, therefore, exhibit relatively low ionic conductivity ($10^{-8} \sim 10^{-7}$ S/cm) [28].

As the ion moves along the polymer chain, old coordination links break and the new connections form. This cation motion is facilitated by the flexing of the polymer chain segments producing a strong coupling between the segmental motion of the polymer and the ionic transport. As the polymer segments are flexible in the amorphous phase, the polymer electrolyte regularly has a higher conductivity in the amorphous phase. Along with that, the anion is relatively weakly bound to the polymer chain, but the flexibility of the polymer chain is the rate also determining for the anion transport [29].

In these electrolytes, the ether oxygen atoms interact with the cations and cause salt solvation. The cation transport is assisted by the segmental motion of the polymer chains. Recognizing the fact that ion conduction takes place in the amorphous phase of polyethylene oxide, considerable researches have been focused on tailoring a flexible host polymer chemical structure with a larger proportion of amorphous phases such as PEO-based block copolymers, star-branched copolymers, and network cross-linked polymers [13].

Various approaches, for instance, the synthesis of branched polymers, copolymers with low molecular weight PEO or the incorporation of plasticizing agents, have been undertaken to decrease the crystallinity and T_g of the electrolyte to enhance the motion of polymer chains that is responsible for improving the ambient temperature conductivity [30].

The most widely used technique to lower the operational temperature of PEO based electrolytes is to add liquid plasticizers, but this gives rise to electrolyte films with poor mechanical properties and higher reactivity towards lithium anode [31].

Few polymer blending methods such as PEO/Polyvinylidene fluoride, PEO/Polyacrylonitrile, PEO/Poly (methyl methacrylate) and PEO/ Polyurethanes have been reported, and the addition of various inorganic fillers has been proposed [32]. However, there is obvious phase separation in the

PEO-based blend electrolytes because the compatibility of PEO and other polymer is not sufficient. Moreover, the mechanical properties and thermal stability of aliphatic polymers are poor, and they cannot be used as electrolytes without additional separator or reinforcing additives. When the third component, such as solid non-ionic plastic crystal material or inorganic fillers of silica, titanium, zeolites, aluminum oxides and others, is introduced into the PEO-based electrolytes to form the composite polymer electrolytes (CPEs), all of the above performances could be, more or less, improved [13].

Therefore, the development of amorphous PEO-based solid polymer electrolytes capable of combining high ionic conductivity with excellent mechanical property is the main goal of the present research.

1.1.5 Introduction to polyethylene glycol

Methyl group capped polyethylene glycol (PEG) with a melting point below room temperature, readily dissolves lithium salts similar to high molecular weight PEO but has better ionic conductivity due to the lower viscosity and higher ionic mobility of lithium ions. As a plasticizer, PEG has been used to improves the electrical properties of solid and gel polymer electrolytes because it provides the ability to decrease the T_g resulting in polymer matrices with increased flexibility. The addition of plasticizers is to effectively improve the ionic conductivity of PEO based SPEs. PEG is also considerably cheaper when compared to commercial high molecular weight PEO [33].

In the research, PEO-based polymer membranes were blended with PEG of a high swelling capability to combine better with PS nanoparticles [34]. High plasticity is beneficial to electrolyte processability, and the composites can be easily manufactured for batteries in various sizes and shapes.

1.1.6 Introduction to LiTFSI

A way to decrease the segment of crystalline phase in the electrolyte is a selection of an appropriate type of anion, which could disturb the regular alignment of polymer chains during crystallization and act like a plasticizer of the polymer matrix. So far, numerous lithium salts have been explored as conducting salts for the PEO-based polymer electrolytes, which primarily include those with weakly coordinating nature, such as hexafluorophosphate (PF_6^-), tetrafluoroborate (BF_4^-), perchlorate (ClO_4^-), triflate ($CF_3SO_3^-$), bis (trifluoromethane sulfonyl) imide ($[N(SO_2CF_3)_2]^-$, TFSI$^-$), bis (pentafluoroethane sulfonyl) imide ($[N(SO_2C_2F_5)_2]^-$, BETI$^-$), and so on.

Among them, $LiPF_6$ is widely used in the electrolytes of LIBs, however, which has relatively poor thermal stability (decomposes at 150℃) and easily forms corrosive HF on exposure to moisture.

On the other hand, LiTFSI has been most extensively studied as a highly dissociative conducting salt for SPEs. This is generally attributed to several intrinsic features of the large anion TFSI$^-$, including (1) the high flexibility of $-SO_2-N-SO_2-$ of TFSI$^-$, being favorable for reducing the crystallinity of PEO matrix (plasticizing effect); (2) the highly delocalized charge distribution of TFSI$^-$, being pivotal for effectively reducing the interactions between Li^+ and TFSI$^-$, thus increasing the dissociation and solubility of LiTFSI in PEO matrix, then and there allows for achieving high values of ionic conductivity [35]; and (3) excellent thermal, chemical and electrochemical stability, being required for stable electrolytes. Attributable to the flexible S-N-S bonds in LiTFSI, the Tg of electrolytes with LiTFSI salt is ordinarily lower than that of electrolytes with other sorts of salts like $LiClO_4$ or $LiCF_3SO_3$. These significant properties of TFSI$^-$ are helpful for developing reliable conductive SPEs for LIBs and other electrochemical devices [27].

Even though LiTFSI-PEO electrolytes show suitable conductivity only at a temperature higher than the melting point (Tm) of PEO, due to the crystallization of PEO, conductivity as low as about 10^{-6} S cm^{-1}, LiTFSI-PEO electrolytes still have other good characteristics, such as light weight, high flexibility and easy for molding.

Also, the concentration of lithium salt in the polymer electrolytes is determined by the molar ratio of $-CH_2CH_2O-$ (EO) /Li$^+$ (i. e., [EO unit] to [Li$^+$]). In the meantime, a molar ratio of ethylene oxide ($-CH_2CH_2O-$, EO) unit/Li$^+$ (i. e., a molar ratio of [EO] to [lithium salt]) of 20 is thus chosen for LiTFSI-PEO, since the concentration of lithium salt for LiTFSI-PEO electrolyte around this ratio region has been found to afford relatively high ionic conductivities at medium-high temperaturess [22].

1.1.7 Introduction to the cross-linked nanoparticles

Practically, an ionic conductivity of about 10^{-3} S cm^{-1} is required to apply solid polymer electrolytes to LIBs. Accordingly, plasticized polymer electrolytes have been recently prepared and they exhibit the ionic conductivities of 10^{-3} S cm^{-1}. Unfortunately, their mechanical properties are weak. In the aim at overcome these problems, the application of cross-linked polymer with good swelling property [36] provides another alternative strategy for the improvement of electrolyte properties that can provide mechanical support while ionic salts and plasticizers can also be contained in the polymer networks [13].

Among them, incorporating cross-linked nano-sized particles into the electrolytes is the better way to improve the membrane strength, rigidity as well as the conductivity, which is attributable to the cross-linked structure and chain extension increases the bonding quality and thus enhances the composite membrane strength [31]. Mechanically stable cross-linked polymer electrolytes countenance safety separation of anodes and cathodes, preventing short-circuits and dissolution of electrode components.

Accordingly, we have prepared three cross-linked polymer electrolytes and reported on the synthesis and characterization, regarding their structural, thermal and morphological features.

1.2　Introduction to lithium-sulfur batteries (LSB)

By contrast with LIB, SPEs due to their physical nature do not display polysulfide shuttling, but exhibit poor ionic conductivities at room temperature, making them unattractive for Li-S batteries [37].

A conventional liquid electrolyte uses lithium bis (trifluoromethyl sulfonyl) imide (LiTFSI) dissolved in ethers, such as dimethoxyethane (DME) and 1,3-dioxolane (DOL) [38]. All essential liquid electrolytes share the following characteristics: a low lattice energy (LLE) lithium salt, a solvent with high relative permittivity, and a strong interaction between the solvent and lithium ions [39][40]. However, two major problems exist in all currently used liquid electrolytes: (i) dissolution of polysulfide (PS) intermediates, and (ii) depletion and oxidation of polysulfides through side reactions with the electrolyte; both issues lead to rapid capacity fading [16].

For a practical Li-S battery electrolyte, the following criteria must be met:

• Ionic conductivity in excess of 10 mS/cm at temperatures ranging from −30℃ to +60℃.

• Electrochemical stability of at least 4V.

• Complete inhibition of the PSS, allowing for 1,000 cycles minimum with very high capacity retention.

- Non-flammable for safety and cost reduction.
- Lithium ion transference number at least 0.6 (essential for high rate capability) [41].

1.2.1 Introduction to LSB solvent systems

Up to now, ethers, sulfones, and fluorinated ethers are the three categories of preferred organic solvents applied in the electrolyte for Li-S batteries. However, a single solvent is rarely used because it cannot meet all of the requirements for battery electrolytes. As for binary or trinary solvent mixtures, different components play specific roles and are complement with each other.

It is well known that lithium cells with carbonate only electrolytes always show notable capacity fading and voltage decline when they work at extremely low temperatures (below 243 K), due to the poor ionic conductivity of electrolyte and low diffusion coefficient of lithium ions in the SEI layer. More strategies should be developed, on the road to achieving electrolytes with wide operating temperatures, such as decreasing EC content and adding co-solvents with low melting point and high boiling point [42]. After modeling and calculating, we have found that nitrogen and lithium ion can make an interaction (Li...N) which is weaker than that of oxygen and lithium ion. Nitrile groups have a high dielectric constant which can increase the polarity of the molecule to increase the solubility of lithium salt. Also, terminated polar groups of the linear molecule can facilitate the flexibility of the linear chain to improve further ioniczation. Moreover, the asymmetric chelation structure may increase the lithium transportation and ionic conductivity [43].

In our group research, two different series of ether based solvents have been synthesized as a potential co-solvent for the electrolyte system: (i) the nitrile terminated ethers can be prepared via the Michael addition reaction of an appropriate alcohol with excess acrylonitrile. The donating ability of the nitrile group is lower than that of the ether group so that the binding force of nitrile groups is weaker than those of ethers, which will allow lithium ions to have much faster ligand exchange in triglyme (G3), leading to faster transport of lithium ions and an increase in ionic conductivity. We used amino/ether-based ligands (AELs) and siloxane core to synthesize a series of nitrile groups terminated polar small molecular compounds to play as liquid electrolyte solvents. (ii) The trifluoromethyl terminated ethers can be prepared via the Williamson ether reaction [44]. New fluorinated solvents are being investigated as nonflammable solvents. A solvent with an F to H ratio >4 appears to have improved thermal properties, fluoro ether may be an alternative in conjunction with cyclic carbonates which could improve the thermal properties of the solvent system [45].

For this part of the research, varied amounts of LiTFSI will be dissolved in a certain weight of different product solvents to make liquid electrolyte solutions. A separator is dipped into the electrolyte solution. The soaked separator is ready to use as a liquid electrolyte. Ionic conductivities, electrochemical windows and thermal properties of each electrolyte solution will be tested.

The research of our designed linear nitrile-ester & fluoro-ester as appropriate co-solvents for carbonate-based electrolytes will be presented in my colleague: Zheng Yue's PH. D dissertation.

1.2.2 Introduction to Polysulfide shuttle

The fully oxidized state (S) and the fully reduced state (Li_2S) are solid, and they are difficult to be dissolved in the electrolyte systems. Nevertheless, lithium polysulfide is highly soluble in the liquid phase. Thus, the electrochemical reaction of the cell is compared with the solid-liquid-solid phase changing, which leads to its unique kinetics [46] [47].

Battery electrolytes should meet the following fundamental requirements:

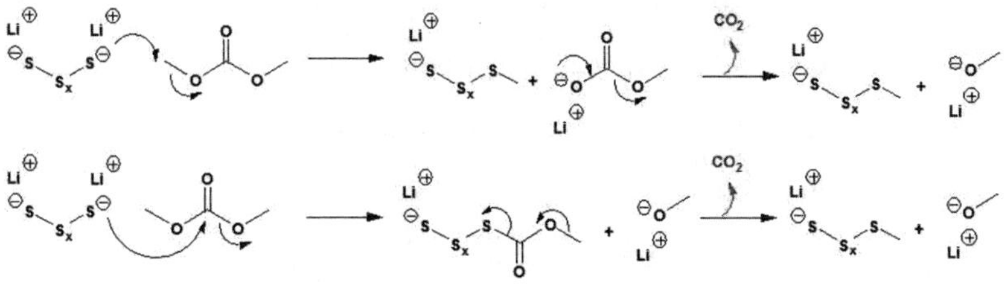

Figure 1.3. Mechanism of polysulfide chain dissolves in DMC

an excellent conductivity for Li^+ cations, which is determined by the diffuse coefficient, chemical, and electrochemical stability; a real affinity for the electrode material so that it could penetrate the matrix and is distributed to the active material. However, traditional electrolytes that meet these requirements are also highly soluble to long chain polysulfide, which causes polysulfide shuttle (PSS) [48].

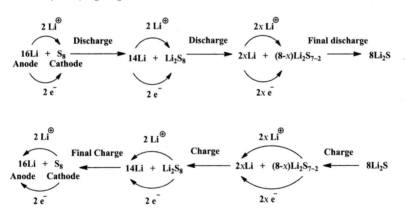

Figure 1.4. Electrochemistry on the cathode of Li-S Batteries

In traditional electrolytes applied in rechargeable Li-ion batteries, S_8, Li_2S_2 and Li_2S have very slight solubility. However, long chain polysulfide, Li_2S_8, Li_2S_6, and Li_2S_4 are highly soluble in most of the organic solvents commonly employed in electrolyte systems. These dissolved sulfur species cause not only capacity fading but also more severe problems. Polysulfide anions diffuse to the anode and be reduced to insoluble Li_2S_2 and Li_2S, lead to self-discharging and deposit on the surface of the anode forming an insulate layer that blocked the transport of Li^+. As a result, the battery undergoes electric capacity fading and decreasing Columbia efficiency. This process is named redox PSS effect.

1.3 Introduction to ionic liquids

Over the last several years, since the discovery of air and water stable ionic liquids (ILs) by Wilkes in 1992 [49], they have been widely promoted as more reliable class of "green solvents" compare with organic carbonates, attribute to their unique properties, such as thermal stability, high electrical conductivity, high polarity [50], and negligible vapor pressures [51], coupled with a wide liquid range. Thus, ILs are recently fascinating considerable attention for application in numerous fields, though they were described almost a century ago.

ILs are a class of designable organic compounds which display two characteristic structures: firstly, they consist only of ions which are poorly. coordinated. Secondly, they are liquids below 100 ℃ [52], if the ILs are liquid at room temperature, they are called room temperature ionic liquids (RTILs) [53]. Combining organic cations with suitable anions allows to tailoring the appropriate physical, chemical and biological properties for ILs, while some even possess unexpected functions resulting from synergetic collaboration between the two components. Hence, the attractive flexibility or 'tunability' in the design has driven phenomenal interest in ILs' synthesis

[54]. The unique structure and performance of ILs as a platform not only offers additional breaks to modify these ionic materials' physical properties (e. g. melting point, density, polarity, viscosity, hydrophobicity/hydrophilicity, solubility) for specific applications, but also offers other gorgeous chemical features such as fundamental ionic conductivity, tremendous thermal, chemical, and electrochemical stability [55].

By either a physical combination of ILs or chemical modification (covalent functionalization or ion exchange metathesis process of the ionic constituents, specific functional groups can be easily incorporated into the ILs skeleton), a number of IL-containing composite materials and functional ILs have been effectively realized and applied in the enormous area.

First of all, nowadays notable efforts for ILs have been made focusing on the design of safer and more environmental kindly solvents [56], due to conventional organic solvents are often toxic, flammable and volatile. Compared with the volatile organic compounds, ILs display excellent dissolution performance for organic, inorganic, and polymer materials, and their immeasurable vapor pressure [57] and non-flammability properties provide them the ability to avoid atmospheric pollution because there would be no loss of solvent through evaporation. The thermal decomposition temperatures higher than 300 ℃ could enhance their recycling efficiency [58]. Accordingly, ILs can be applied to replacing traditional volatile organic solvents for a host of practices linked to green chemistry and clean technology such as organic reactions [59], extraction, catalysis, and separation processes [55].

Secondly, ILs have a very broad range of viscosities, which may vary between 20 and 40,000 cp as are compared with the viscosities of typical organic solvents which are in a range of 0.2 and 100 cp. For ILs, only minor change of structure may result in a significant change in viscosity. Thus, ILs are discovered extensive use as engineering fluids or as innovative lubricating systems [50]. IL-based materials could straightforwardly be operated under extreme conditions such as high or low temperature and high vacuum or pressure, owing to their significant thermal stability in a quite wide liquid phase temperature range from 200 to 300 ℃, with designable excellent mechanical, chemical, and electrochemical stability.

Thirdly, the special designable structures and functional properties not only retain the key features of the original materials but also possess the characteristics of ILs. It is thought that the ionic nature of ILs, which the molecular materials lack, can offer these materials with intrinsic ion conductivity as well as a substantial ionic skeleton for producing progressive materials. At the meantime, ILs exhibit perfect electrochemical stability in a wide electrochemical window of 2 ~ 5 V [60]. Thus, ILs, especially RTILs, would be an excellent candidate for prospective applications as encouraging electrolyte bases or additives in lithium secondary batteries and other energy related applications, such as fuel cells, supercapacitors [61], and dye sensitized solar cells [62] [63].

Since many factors can affect their conductivity, such as viscosity, density, ion size, anionic charge delocalization, aggregations and ionic motions, ILs are expected as key materials which might give a solution to the safety problems of batteries due to their non-flammable property. However, the selectivity of carrier ion transport is one of the problems when ILs are applied to batteries. Since ILs do not include electroactive species, it is necessary to add salts or acids before their use. The mobility of electroactive ions shows great effects on the electrical power of the batteries as lithium cation does on lithium-ion batteries, while the introduction of ethylene oxide chains attached to the methyl pyrrolidine, butyl imidazolium ring or diethylsulfide may lead to lower glass transition temperature, and also improve ionic conductivity [41].

In these applications, ILs are constantly discussed as beneficial and high potential building blocks for innovative advanced functional materials. The resulting ILs are expected to offer many more advantages over the traditional molecular materials.

1.3.1 Introduction of cations and anions used in ILs

The wide range of possible cation and anion combinations allows for a large variety of tunable interactions and applications [64]. Generally studied ILs are consist of bulky, N-containing organic cations in combination with anions, alternating from the inorganic ions to more various organic species. Simple deviations in the cation and anion groupings or the nature of the moieties attached to each ion allows the properties of ILs to be tailored for specific applications.

The cation and its structure can positively influence the physical properties as well as interact via dipolar, ð-ð, and n-ð interactions with dissolved molecules. However, its range of effects has not been studied as extensively. [65] Generally, the commonly used cations include alkyl imidazolium [RR' Im]$^+$, alkyl pyridinium [RPy]$^+$, tetraalkylammonium [NR$_4$]$^+$, tetra alkyl phosphonium [PR$_4$]$^+$ and others [62]:

Most widely used anions consist of chloride Cl$^-$, bromide Br$^-$, iodide I$^-$, hexafluorophosphate [PF$_6$]$^-$, tetrafluoroborate [BF$_4$]$^-$, nitrate [NO$_3$]$^-$, methanesulfonate (mesylate) [CH$_3$SO$_3$]$^-$, trifluoromethane sulfonate (triflate) [CF$_3$SO$_3$]$^-$, bis-(trifluoromethane sulfonyl) amide [CF$_3$SO$_2$)$_2$N]$^-$ or abbreviated from here on as [TFSI]$^-$, dicyanamide [DCA]$^-$, and others [62].

1.3.2 Introduction of DILs

As a new family of ILs, DILs are consist of the anion and the doubly charged cation which is composed of two singly charged cations as head groups linked by a variable length of alkyl or oligo ethylene glycol chain as a rigid or flexible spacer. Thus, the DILs matrix created offers the opportunity to investigate (a) the influence of cation and anion variation, (b) the influence of the chain length.

Recently, the number of DILs described in the literature is growing rapidly. Anderson et al. presented the synthesis and characterisation of a variety of DILs containing alkane spacers [65]. Kubisa and Biedron investigated DILs obtained by the functionalization of PEG with triphenylphosphine [66]. Ohno and co-workers have shown that dicationic PEG-based molten salts are bearing imidazolium cations as ion conducting materials [67]. Dicationic and polycationic ILs linked by alkyl chains display extraordinarily higher chemical and thermal stability than their mono-cationic analogs published by Armstrong and coworkers [65]. It has also been presented that the acute toxicity of DILs is in several cases below the levels detected for those mono-cationic and that the use of head groups connected via polyethylene glycol could be identified as structural elements reducing the toxicity [50].

Based on the literature, DILs possess distinctive features in critical micelle concentration [57], such as a wider liquid range and higher thermal stability, better behavior as electrolytes characteristics, and higher thermal stabilities than mono-cationic ILs and other traditional solvents [68]. As we know, ILs should have the capability to dissolve more Li salt to have a higher Li^+ conductivity, if ILs are intended to be applied in Li batteries. Thus, by incorporating a PEO oligomer into the DILs structure, the conductivity could be enhanced primarily by improving cation transport in the PEO segment, although the viscosity of DILs does not decrease [69] due to the weak interaction of ethylene oxide segment compared to the alkyl chain [70].

Until now most of the researchers have focused on mono cationic type ILs, although the number of DILs described in the literature is increasing [62]. For DILs, the relationship studies between their structure and physicochemical characteristics and molecular structures are, as yet, still relatively rare [71]. Therefore, it is urgent and necessary to explore other new DILs structures to gain further understanding and extend the applications of DILs as electrolyte components.

1.3.3 Introduction of fluorinated room temperature ionic liquids (FRTILs)

Although many ILs have been reported, most are not suitable for electrochemical applications because of poor conductivity (<1 mS) and/or high viscosity (>100 cP) at room temperature, also due to a poor electrochemical window. However, the combination of a fluorinated cation with an anion such as TFSI or DCA has resulted in ILs that display non-volatility, non-flammability [72], relatively wide electrochemical windows and also low viscosities and high conductivities that are favorable for electrochemical applications [73].

Among the tools available to synthetic chemists to tune ILs properties, selective fluorination is among the most productive, having been extensively developed and explored since the introduction of diversely located fluoro substituents including terminal chain, linking arm or core position which provides an excellent opportunity for investigating the relationship between structures and properties and for modifying and optimizing the physical/chemical properties of compounds [74].

Meanwhile, although many FRTILs have also been synthesized and shown to display excellent modified properties due to the existence of fluoro substituents, investigations involving this kind of RTILs bearing fluoro substituents have been reported only rarely. In this work, we have described the design and syntheses of twelve novel unsymmetrical mono cationic based FR-TILs with side chains. Their properties and potential applications as solvents, electrolytes, high ion transport materials and templates have also been extensively developed.

1.4 Introduction to measurements

1.4.1 Scanning electron microscopy

The morphology of the polymer nanoparticles was analyzed via a scanning electron microscopy (SEM) (Cambridge, Leica) with an accelerating voltage value equal to 15 kV. The samples, analyzed in their cross section, were coated with a thin gold layer by means of a sputter coating (Polaron, model SC502 sputter coater) [21].

1.4.2 Differential scanning calorimetry (DSC)

DSC is one of the most convenient approaches to throw light on the degree of crystallinity along with the thermal behavior of any polymeric samples [4]. The glass transition temperature (T_g) of the resulting polymer film was determined via DSC measurement. The T_g of a polymer indicates the transition from a rubbery into a glassy state. Therefore, the polymer is flexible above the T_g and hard and brittle below the T_g [21].

In our study, DSC analysis was carried out with a Mettler Differential Scanning Calorimeter instrument (Figure 1.5.) under argon purge atmosphere flow (70.0 mL/min) [75] between $-100\ ℃$ and $150\ ℃$ along with liquid nitrogen as a cooling element [34]. Samples around 5~7mg were sealed in 40 μL aluminum pans with perforated lids to allow the release and removal of the decomposition products [20]. Two segments of DSC scanning included:

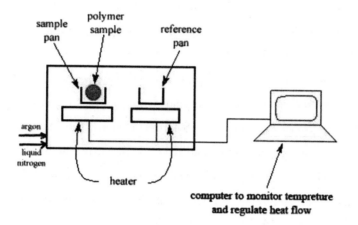

Figure 1.5. A schematic of DSC experiment

1) $-100\ °C$ isothermal for 10.0 min, 2) with a heating rate of 10 °C/min and the temperature range from -100 to 150 °C. Another empty aluminum pan was placed as a reference.

1.4.3 Thermogravimetric analysis

The thermal stability of the polymer electrolytes was measured by thermogravimetric analysis (TGA). Thermogravimetric measurement was carried out with a TGA/SDTA851e thermal analyzer. Around 5.442 mg of sample were placed in a TGA aluminum pan. The sample was heated from 25 °C to 130 °C for 15 minutes and cooled rapidly to 25 °C for 15 min. Then the samples were heated up to 400 °C at a rate of 10 °C/min [12]. The fully amorphous nature of the polymer was observed under thermal characterization.

1.4.4 Cyclic voltammetry

The electrochemical stabilities of the SPE membranes were determined by cyclic voltammetry (CV) using a potentiostat/ galvanostat (Solartron impedance analyzer). A testing cell was assembled to determine the oxidation potential with stainless steel as working electrode and lithium foil serving as a counter and reference electrode. All the scans were made at 25 ℃ at a scan rate of 10 mV/s within a voltage range of 0 V to 5.0 V versus Li/Li$^+$ [76].

1.4.5 Ionic conductivity

The ionic conductivity of the obtained compounds was measured by the complex impedance method using an impedance analyzer (Solartron model SI-1287; Schlumberger) coupled to a Solartron model-1260 frequency response analyzer. The membranes were sandwiched between two stainless steel electrodes of 2.54 cm^2 in the area for conductivity measurement [77]. All the measurement was performed in a glove box filled with a dry nitrogen gas in a temperature range of 25 – 70 ℃ [78].

The ionic conductivity (δ, in S cm^{-1}) was plagiaristic from the value of resistance (R, in Ω) of the bulk electrolyte obtained in the complex impedance diagram, in relation to the equation: $\delta = L/(RS)$, where L (cm) is the thickness of the membrane of the polymer electrolyte, and S (cm^2) is the area of the membrane [22].

1.5 New Electrolytes in this thesis

In this thesis, as mentioned above, the area of electrolytes is chosen. Different components of solid and liquid electrolyte systems and their advantages and disadvantages will be discussed. The major focus of the work is to develop a better electrolyte system for rechargeable lithium batteries.

In lithium-ion battery chemistry, the important issues for SPE systems are associated with the polymer matrix. Although PEO is undoubtedly the best system still on hand for the base matrix in SPEs, the high crystallinity of PEO matrix is clearly not suitable for ambient temperature ionic conductivity at the desired level. On the way to overcome this difficulty, a series of novel lithium salts were synthesized and incorporated into a conventional PEO based SPE system, include three novel NPLS and four novel fluorinated dilithium salts with low lattice energy, and excellent thermal stability was presented. The ionic conductivity of these PEO-based SPEs was observed noteworthy, due to the decrease in the glass transition temperature of the polymer. These jobs are shown in Chapters 2 and 3.

Another main problem of the current liquid electrolyte system is the safe concern due to the existing of volatile and flammable organic carbonates, and poor low-temperature performance. A novel electrolyte solvent system should present an excellent ionic conductivity without volatile solvents. Varieties of non-volatile polar solvents with very low freezing point should be designed and studied for the liquid electrolyte systems. The approach involves the strategic design and synthesis of a series of RTILs to address the PSS problem. i) TFSI-based ILs are non-flammable, ease of synthesis, excellent thermal stability, but in general, cannot dissolve enough lithium salt (typically $<0.5M$) to form the relatively high concentration. ii) DCA-based ILs have much lower viscosity than TFSI-based ILs and tend to display much better ionic conductivity, by using DCA as the counter anion for the ILs. DCA anion is usually unusable for LIB applications (4V electrochemical window), but can perhaps be used for LSB as a result of the much lower operating voltage. iii) Asymmetric fluoro-based RTILs may exhibit good ionic conductivity, excellent electrochemical & thermal stabilities, but need further measurements. Higher lithium ion density, excellent thermal property, and lower cost will demand the development of new RTILs. This area is discussed in Chapters 4 & 5.

Thus, the goal of this thesis will be to create new electrolyte systems, which would meet the demands of automotive manufacturers, leading to the commercial production of rechargeable lithium batteries.

Chapter 2   Active Nano-Particles Filled SPEs

2.1   Introduction

Pressure to reduce carbon emissions, global warming and over reliance on fossil fuels are a few of the factors that make electric and hybrid electric vehicles (EV/HEV) more attractive alternatives to the burning of fossil fuels. Lithium-ion (Li-ion) batteries represent a much more advanced alternative to lead-acid batteries since they have a much higher power to weight ratio (2.59 MJ/Kg) [2] [3]. Moreover, Li-ion batteries possess favorable versatility in many other applications, e. g., in computers, cell phones, cameras, camcorders, and medical devices. Nevertheless, current Li-ion batteries have limitations that must be overcome before they can be used on a mass scale. Three of the most important challenges are power density, cost, and safety of operation. Electrolytes are the second most expensive (after the cathode) component of Li-ion batteries. Battery companies are desperately seeking low cost, safer, and high performance electrolytes.

Current state-of-the-art Li-ion batteries employing liquid electrolytes consists of two components: a Li-ion source and an organic solvent. The Li-ion source is generally a single or mixed fluorinated Li-salt [e. g. , $LiPF_6$, CF_3SO_3Li, and $(CF_3SO_2)_2NLi$]. A commonly used organic solvent is ethylene carbonate (EC) because of its low cost, good electrochemical stability, and high dielectric constant, which facilitates dissociation of the Li-ion source, leading to high ionic conductivities. Other carbonates, such as dimethyl carbonate (DMC) and propylene carbonate (PC) are often used in conjunction with EC to reduce viscosity, as well as to increase the wettability of the electrolytic solution with battery components, e. g. , separator and electrodes. Despite these significant attributes of carbonate solvents, they are not recommended for EV/HEV batteries because of their highly flammable characteristics in the absence of thermal/flame protection devices. A small Li-ion battery requires ~ 5 mL of liquid electrolyte, while an EV battery, which contains many small batteries connected in series, requires anywhere from 500 – 1000 mL of electrolyte. Thus, liquid electrolytes based full size batteries can potentially be explosive under abusive conditions (e. g. , shorting, crushing, or excessive overcharging) and occasionally, under normal conditions (e. g. , over discharge, resistive and/or forced over discharge), because the batteries may undergo thermal runaway that generates a sharp rise in temperature and results in serious hazards of fire and explosion [21]. Accordingly, there is a compelling need for strategic design and development of an advanced solvent free electrolyte system, which is free of leakage and possesses high ionic conductivity and desired electrochemical and mechanical properties.

The availability of lithium rechargeable batteries featuring solvent-free highly conductive solid polymer electrolyte (SPE) systems will have a major impact on the EV/HEV industry, leading to a significant reduction in environmental pollution and improved performance, compared with current carbonate based liquid electrolytes. The driving force for this advance is the strong potential for achieving high energy densities, high cell voltage, and superior self-discharge characteristics, while largely mitigating deficiencies, such as leakage, instability, and difficulty in the manufacture of the major flat types are currently prevalent in Li-ion batteries containing liquid electrolytes.

The electrochemical properties of SPEs required for EV/HEV applications include: ionic conductivity greater than 10^{-3} S/cm at room temperature, electrical conductivity less than 10^{-7} S/cm, electrical breakdown at greater than 5 V/m, and stability of the electrolyte adjacent to cathode material up to at least 4.5 V versus lithium. Other desired features include high mechanical strength, glass transition temperature (T_g) much lower than room temperature, thin film processability, good interface properties (compatibility and adhesion), and smooth operation at ambient temperature. Moreover, the system must be amenable to mass production at reasonable cost.

At this time, a serious drawback of all SPEs is their inadequate conductivities (much lower than 10^{-3} S/cm) at ambient temperatures. Good ionic conductivity is essential to ensure that a battery system is capable of delivering usable amounts of power at a high rate, a critical requirement for EV/HEV batteries. It has been predicted that an SPE possessing room temperature conductivity near 10^{-3} S/cm would lead to the mass scale production of long awaited and significantly safer, high energy density batteries [46]. Among the solvent free polymer electrolyte systems that have been the most investigated in past decades is polyethylene oxide (PEO). The main advantages of PEO as a host are its chemical, mechanical and electrochemical stabilities since it contains only strong unstrained C-O, C-C, and C-H bonds. PEO is very flexible ($T_g = -61°C$) because of the presence of swivel ether linkages and the repeat unit ($-CH_2CH_2O-$) providing just the right spacing for maximum dissolution of lithium salts. Owing to the presence of sufficient interchain entanglement, PEO electrolyte behaves like a rubbery material but contains both crystalline and amorphous regions. It should be noted that lithium ion conduction takes place in the amorphous phase via diffusion, which occurs through a complex mechanism involving the PEO segmental mobility [28] (Figure 1.2). PEO electrolytes also exhibit excellent melt processing capability which is very desirable for the scale mass production of batteries.

Despite these positive attributes, all PEO based electrolytes exhibit poor room temperature ionic conductivity ($< 10^{-5}$ S/cm), because the degree of crystallinity in a PEO system increases with an increase in concentrations of lithium salts, leading to a marked decrease in ionic conductivity before an acceptable value is attained. As a result, the potential of the PEO system will remain impractical unless the room temperature conductivity is increased from $< 10^{-5}$ S/cm to the "magic number" of $> 10^{-3}$ S/cm.

Recently, there have been many attempts to improve ionic conductivity of PEO electrolytes with regard to (i) the minimization of the crystallinity of PEO to make a fully amorphous polymer [30], (ii) the investigation of the effect of mixing nano sized inorganic filler on the ionic conductivity of PEO composite [32], and (iii) the addition of liquid plasticizer (both low and high molecular weight) to prepare PEO electrolyte [31]. Unfortunately, these approaches have yet to achieve anywhere near the magic number. Even though extensive work has been done with other low T_g based SPE systems (viz. polyphosphazene and polysiloxanes), they are far behind the current PEO systems, because of the low dissolution of lithium salts in those polymer matrices [33]. Consequently, finding the ideal SPEs having good transport properties still gives the impression to be a challenging task.

Researchers can improve the performance of the polymer electrolyte by reducing the crystallinity of the polymer, increasing the concentration of ions and the proportion of the amorphous region contained in the system [14]. Along with decreasing glass transition temperature (T_g) of the polymer electrolyte system and improving the capacity of lithium ion dissociation [21], the methods consist of blocking, grafting, crosslinking, compositing and blending various polymers materials [23].

Accordingly, investigation of new polymeric salts is essential, although several factors are still needed to be considered while making a choice for the best lithium salt, and the key components include performance, price, and safety. The performance of salt is related to its conductivity at various temperatures, thermal and electrochemical stability.

Based on these, this paper examines the performance of three cross-linked polystyrene lithium salts (PSLSs): PSTFSILi, PSPhSILi, and PSDT-TOLi, which belong to the category of low lattice energy lithium salts. To obtain a balance for the compatibility of ionic conductivity and mechanical properties, this series of polymer salts are blended with high molecular weight PEO (Mw = 4×10^6 g/mol), PEG (Mw = 1000 g/mol) with high swelling capability, complexed with LiTFSI [34].

The most important aspect which has prompted us to study this class of SPEs is that the current state-of-the-art liquid electrolytes exhibits very low cationic transference numbers ($t^+ = 0.2 - 0.3$), which causes polarization of the electrolyte, leading to an increase in resistivity, especially near or below subzero temperatures [6]. Recharging the cell then requires more time, energy, and electrochemical potential. In order to circumvent this problem, it is essential that, in addition to high conductivity, an electrolyte exhibits a high cationic transference number (> 0.6) for smooth and efficient charging and discharging characteristics. Since the anionic part of the lithium salt is immobile (covalently linked to crosslinked polystyrene microparticles) in the proposed SPEs, we anticipate that these SPEs will behave similarly to a single-ion conducting polymer, which exhibits a higher cationic transference number, leading to the efficient charging and discharging. Addressing the issue of cationic transference number is significantly important for EV/HEV application since these vehicles are expected to survive extreme weather conditions. In order to accomplish the aforementioned goals, we present herein a new synthetic strategy to develop a commercially viable product.

In choosing this system, PEG has been used as a plasticizer to improve the electrical properties of solid polymer electrolytes since it provides the ability to decrease the T_g resulting in polymer matrices with increased flexibility. The addition of plasticizers is to effectively improve the ionic conductivity of PEO based SPEs [30]. PEG is also considerably cheaper compared to commercial high molecular weight PEO [33]. Among that, incorporating cross-linked nano-sized particles with good swelling property [36] into the electrolytes is the better way to improve the membrane strength, rigidity as well as the conductivity [13]. Mechanically stable cross-linked polymer electrolyte membranes are also able to safety separation of anodes and cathodes, preventing short circuits and dissolution of electrode components [31].

The primary goal of our study is to identify new, stable, and environmentally friendly PSLSs which exhibit excellent electrochemical and thermal properties for potential use in Li-ion batteries.

2.2 Experimental

2.2.1 Materials

Styrene (St), divinyl benzene (DVB), ammonium persulfate (APS), triton X – 100 (emulsifier), sodium dodecyl sulfonate (SDS), chlorosulfonic acid, lithium hydroxide monohydrate anhydrous, 1, 3-dithiane, trifluoromethane sulfonamide, benzenesulfonamide, N-Butyllithium solution, hydrogen peroxide, triethyl orthoformate, 1, 2-ethanedithiol and acetic acid were purchased from Sigma-Aldrich Chemical Co., Ltd (USA). Acetone, acetonitrile, dichloromethane and methanol were supplied by Fisher Scientific Co., Ltd., (USA). All other reagents for preparing the compounds were commercial products.

2.2.2 General procedures for the synthesis of PSTFSILi

2.2.2.1 Synthesis of cross-linked nanoparticle polystyrene (XLNP). Styrene is hard to store in pure form. The commercial product usually contains added inhibitors (such as a trace of hydroquinone). The material was washed with 10% NaOH solution to remove inhibitors (tert-butyl catechol), then washed with water two times, and dried for several hours with $MgSO_4$. Methylene chloride was used as a solvent to separate the pure styrene through column chromatography. The solvent was removed via high vacuum for 0.5 hr. The pure styrene was stored at 0 ℃.

Chloroform was added to vinyl anisole, and the mixture was washed with 10% NaOH, the yellow organic material was washed twice with water, then dried over $MgSO_4$ and isolated by vacuuming distillation for 0.5 hr, pure vinyl anisole was stored at 0 ℃.

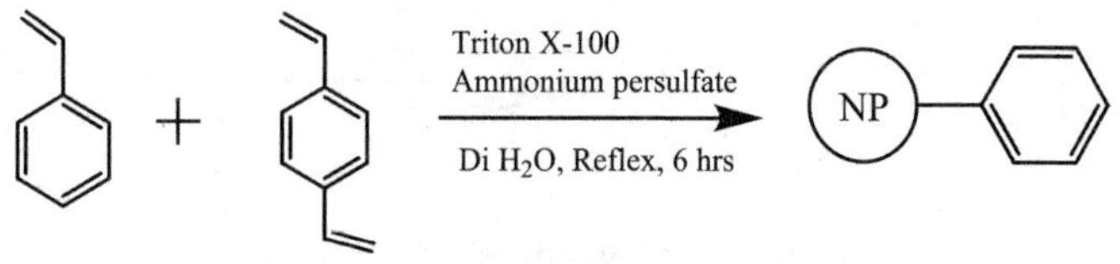

Scheme 2.1. Synthesis of cross-linked nanoparticle polystyrene (XLNP)

Distilled water was degassed under argon for 3 hr. Then 5.5 ml of pure styrene and 0.22 ml of 1,4-divinylbenzene (DVB) were mixed thoroughly with argon, and 60 ml of degassed water was added to it, followed by 0.04 g ammonium persulfate. The mixture was stirred for 5 min after which 0.4 g of sodium dodecyl sulfate and 0.4 g of Triton X – 100 were added with stirring for 1 hr. The mixture was heated for 6 hrs at 75℃ – 85℃ then poured onto a glass plate and allowed to stand for two days to obtain the pale yellow crystals. The crystals were ground into powders, then sonicated for 15 min with 15 ml of methanol at 40 ℃ [79]. The mixture was centrifuged for 30 min. The procedure was repeated three times to obtain the white product XLNP. (Yield: 4.2 g, 81.2%).

FT – IR: 2900 – 2950 cm^{-1}, 2800 – 3100 cm^{-1}, 1601.97 cm^{-1}, 1450 – 1500 cm^{-1}, 756 cm^{-1}, 697 cm^{-1}

2.2.2.2 Chlorosulfonation of cross-linked nanoparticle polystyrene (XLNP-SO_2Cl). PS of 4.30g and 40ml of dichloromethane were placed in a flask and stirred overnight for solvent absorption. A mixture of 12.5 ml of nitromethane and 13 ml of chlorosulfuric acid was added drop by drop to it. The reaction mixture was heated for 7 hr at 40℃. After isolating the crude product, 10 ml of dichloromethane was added to it. The solution was filtered with a sintered funnel, washed twice with 10 ml of acetonitrile, and then washed twice with 10 ml of acetone. The solvents were removed under vacuum at 70 ℃ overnight [80]. (Yield: 7.99g)

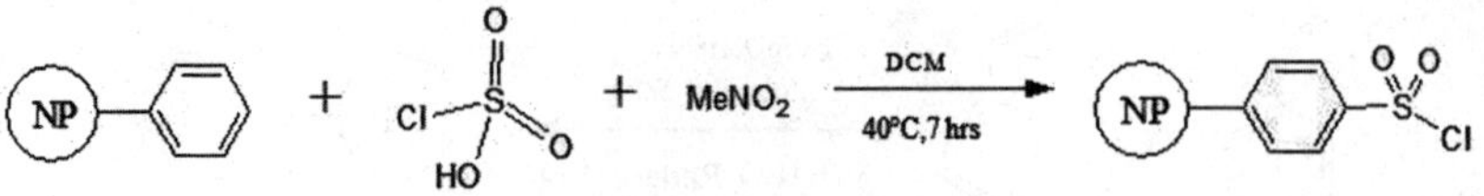

Scheme 2.2. Synthesis of XLNP-SO$_2$Cl

FT – IR: 2928.05 cm^{-1}, 1365 ± 5 (as) & 1180 ± 10 (s) cm^{-1}, 1160 – 1140 (s) & 1350 – 1300 (s) cm^{-1}, 1450 – 1500 cm^{-1}, 772.50 cm^{-1}, 672.09 cm^{-1}.

2.2.2.3 Synthesis of PSTFSILi. PS – SO$_2$Cl (2.50g, 12.2mmol), trifluoromethanesulfonamide (1.97g, 13.2mmol), lithium hydroxide monohydrate (1.11g, 26.5mmol) were placed into a 50ml round-bottomed flask, and 30 ml of anhydrous acetonitrile was added to it. The mixture was stirred at room temperature overnight and then heated to 50 ℃ for 2 hours and cooled down. The solution was sonicated and centrifuged after washing it with 15 ml methanol, and the process was repeated twice with water and acetone washes. The material was placed in a high vacuum at 70 ℃ overnight. (Yield: 2.55g)

FT – IR: 2900 – 2950 cm^{-1} (m), 1350 – 1300 (s) cm^{-1} & 1180 – 1140 (s), 1000 – 1400 cm^{-1}, 1639.23 cm^{-1} (sh), 795.44 cm^{-1} (m), 677.52 cm^{-1} (m).

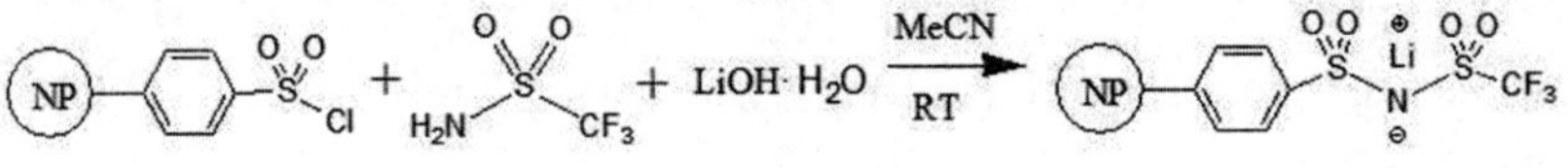

Scheme 2.3. Synthesis of PSTFSILi

2.2.3 Synthesis of PSPhSILi

PS – SO$_2$Cl (2.50 g, 12.3 mmol), benzenesulfonamide (1.94 g, 12.3 mmol), and lithium hydroxide monohydrate (1.04 g, 24.7 mmol) were placed in a 50 ml round-bottomed flask, and 30 ml of anhydrous acetonitrile was added to it. The mixture was stirred at room temperature overnight and then heated to 50 ℃ for 2 hours and cooled. The solution was sonicated and centrifuged after washing it with 15 ml of methanol centrifugation, and the process repeated with water and acetone washes. The material was placed in a high vacuum at 70 ℃ overnight. (Yield: 2.55 g)

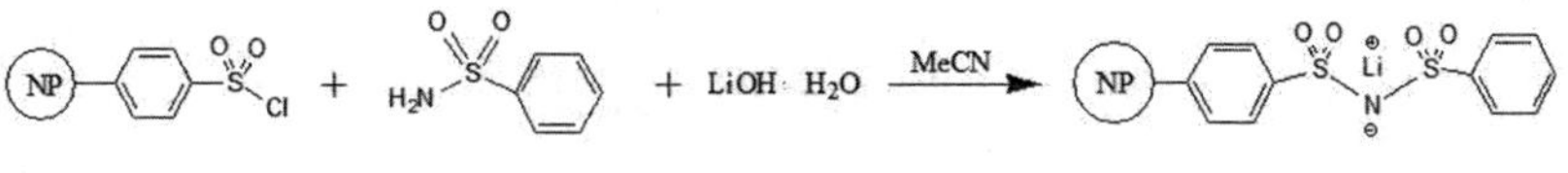

Scheme. 2.4. Synthesis of PSPhSILi

FT – IR: 3063.53 cm^{-1} (s), 2850 – 2950 cm^{-1} (sh), 1400 – 15000cm^{-1} (s), 1000 – 1200 cm^{-1} (s), 1638.34 cm^{-1} (sh), 1600.48 cm^{-1} (m), 832.79 cm^{-1} (m), 678.43 cm^{-1} (m), 580.10 cm^{-1} (m).

2.2.4 Synthesis of PSDTTOLi

2.2.4.1 Synthesis of polystyrenesulfonyl – 1, 3 – dithiane. 1, 3 – Dithiane (0.32 g) was placed in a 5ml 3 – neck round bottomed flask. Then 1.8 ml of 2.5 M n-butyllithium solution in hexane was added to the flask under Argon. The mixture was stirred at 0 ℃ for 1 hour, then to which PS – SO$_2$Cl (0.50 g) was added which was soaked with THF for 3 hours. The reaction was continued overnight. The mixture was sonicated and filtered after washing it with 15 ml methanol. The product was dried in a high vacuum at 70 ℃ overnight. (Yield: 0.51 g)

FT – IR: 2900 – 2950 cm^{-1}, 1638.61 cm^{-1}, 1595.89cm^{-1}, 1350 – 1495 (s) cm^{-1}, 1000 – 1180 cm^{-1}, 831.89 cm^{-1}, 775.60 cm^{-1}, 673.52 cm^{-1}, 578.62 cm^{-1}.

2.2.4.2 Synthesis of polystyrenesulfonyl – 1, 3 – dithiane – 1, 1, 3,

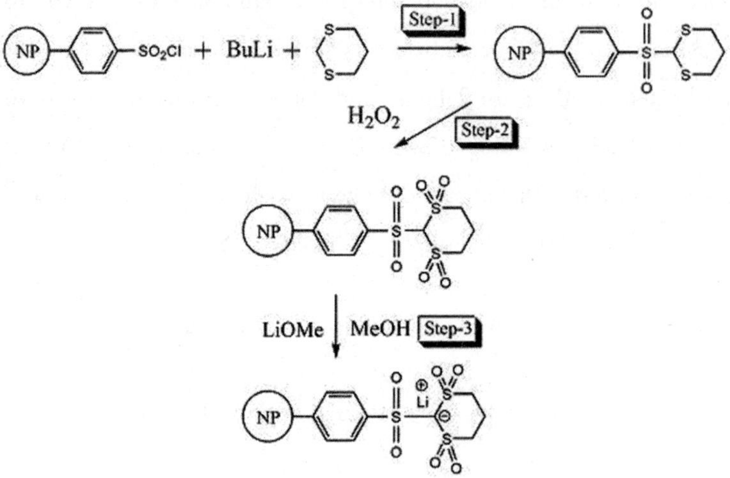

基于新型可充电锂电池固态和液态电解质的研究 >>>

Scheme. 2.5. Synthetic Routes of PSDTTOLi

3 – tetraoxide (PSDTTO). Polystyrenesulfonyl – 1, 3 – dithiane (0.50 g) was placed in a 50ml round bottomed flask. Then 15 ml of acetic acid as a solvent was added to it. Hydrogen peroxide (10 ml) was added to the flask. The mixture was heated and stirred at 60℃. The reaction was conducted for three days with the further addition of 1ml of hydrogen peroxide/day. The product was filtered and washed with 15 ml of water. The product was placed in a high vacuum at 70℃ overnight. (Yield: 0.34 g)

FT – IR: 2900 – 2950 cm^{-1}, 1717.43 cm^{-1}, 1639.34 cm^{-1}, 1600.46 cm^{-1}, 1350 – 1495 (s) cm^{-1}, 1000 – 1225 cm^{-1}, 831.89 cm^{-1}, 775.61 cm^{-1}, 673.52 cm^{-1}, 578.62 cm^{-1}.

2.2.4.3 Synthesis of PSDTTOLi. Polystyrenesulfonyl – 1, 3 – dithiane – 1, 1, 3, 3 – tetraoxide (0.34 g), and lithium methoxide (0.08 g) were placed into a 50 ml round bottomed flask, and 20 ml of methanol was added to it. The mixture was stirred at room temperature for two days. The product was washed with methanol twice followed by with acetone. The product was dried in a high vacuum at 70℃ overnight. (Yield: 0.35 g)

FT - IR: 2900 - 2950 cm^{-1}, 1705.73 cm^{-1}, 1637.34 cm^{-1}, 1600.75 cm^{-1}, 1410 - 1495 (s) cm^{-1}, 1000 - 1190 cm^{-1}, 832.79 cm^{-1}, 776.25 cm^{-1}, 678.43 cm^{-1}, 582.81 cm^{-1}.

2.3 Results and Discussion

Solid polymer electrolytes (SPEs) comprising of cross-linked polystyrene nanoparticles poly (styrene trifluoromethane sulphonyl imide of lithium) (PSTFSILi), high molecular weight poly (ethylene oxide) (PEO, Mw = 4x10^6 g/mol), poly (ethylene glycol dimethyl ether) (PEG, Mw = 1000 g/mol) complexed with lithium bis (fluoro sulfonyl) imide (LiTFSI), have been prepared and characterized. Moreover, another two analogous polymer salts: poly (styrene benzene sulphonyl imide of lithium) (PSPhSILi) and poly (styrenesulfonyl - 1, 3 - dithiane - 1, 1, 3, 3 - tetraoxide of lithium) (PSDTTOLi) were comparatively studied. The polymers were characterized by FTIR spectroscopy, EDS, SEM, and TGA and tested for possible application as an electrolyte for lithium-ion batteries. Their physicochemical properties have been systematically investigated regarding thermal stability, electrochemical stability, and ionic conductivity. The cross-linked nano-sized polymer salts markedly increase film strength and decrease the glass transition temperature (T_g) of the polymer electrolyte membranes. Meanwhile, the enhancement and the improvement in the ionic conductivity and thermal stability have correspondingly been observed. The conductivity was found to be optimum in the quantity of 7.52×10^{-05} S/cm at room temperature and 3.0×10^{-03} S/cm at 70 ℃ for PSTFSILi film, while PSDTTOLi film showed the even better performance of 1.54×10^{-04} S/cm at room temperature and 3.23×10^{-03} S/cm at 70 ℃.

2.3.1 Characterization

We synthesized three cross-linked polymer lithium salts and reported on the synthesis and characterization, regarding their structural, thermal and morphological features.

2.3.1.1 FTIR. The FTIR spectra of the nanoparticles were performed with potassium bromide. For PSTFSILi, the peak at 1000 – 1400 cm^{-1} corresponds to the stretching of C-F bond. The broad peaks in a range of 1350 – 1300 cm^{-1} (s) & 1180 – 1140 cm^{-1} (s) are due to-SO_2N-bond.

For PSPhSILi, stronger peaks around 3063.53 cm^{-1} correspond to aromatic C-H stretching and 1500 – 1400 cm^{-1} relate to aromatic C = C stretch in-ring aromatics. With the encouraging results on the sharp peak around 1639 cm^{-1} is the most prominent peak, which indicates that this reaction is rather successful. The peak around 1630 cm^{-1} represents the S-N-S combination bond, which shows the reaction is completed.

For PSDTTOLi, after lithiation of PSDTTO, the peaks at 1717.43 cm^{-1} shift to 1705.73 cm^{-1}, and 1639.34 cm^{-1} shift to 1637.34 cm^{-1}, which was assigned to the stretching vibration band of carbon with the existence of three electron withdrawing groups ($-SO_2$).

The illustration of the cross-linked structures is provided in Figure 2.1. Polystyrene backbone serves as a medium to soak or swells more polymeric matrixes; divinylbenzene is engaged as a cross-linker, which could increase the chain extension the bonding quality, the functional group plays a role as the ionic conductor. Thus incorporating the cross-linked nano-sized particles into the electrolytes is the better way to improve the membrane strength, rigidity as well as the conductivity.

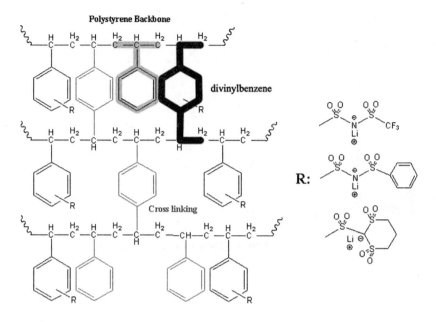

Figure 2. 1. Schematic illustration of cross-linked PSLSs

2. 3. 1. 2 Energy Dispersive Spectrometer. The Scanning Electron Microscope is equipped with an Energy Dispersive Spectrometer (EDS), which provides chemical analysis of the field of view or spot analyses of minute particles. Figs. 2. 2 – 2. 4 show the EDS elemental analysis of each sample,

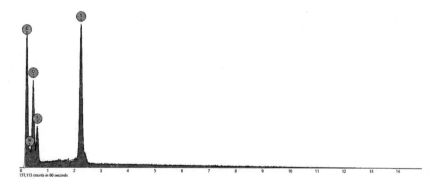

Figure 2. 2. EDS for PSTFSILi nanoparticle

while the obtained elemental concentrations from EDS studies are presented in Table 2.1.

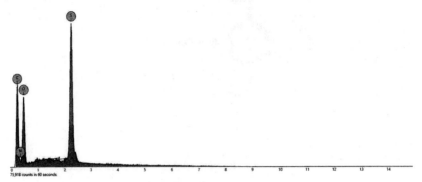

Figure 2.3. EDS for PSPhSILi nanoparticle

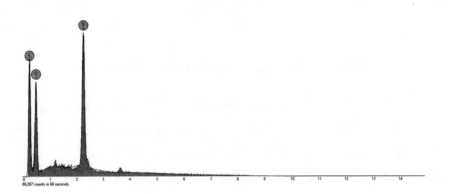

Figure 2.4. EDS for PSDTTOLi nanoparticle

Table 2.1. Elemental concentrations determined using Energy Dispersive Spectroscopy

Sample	At. % C	At % N	At. % O	At. % S	At. % F	Percent Substitution
PSTFSILi	48.9	13.5	26.4	3.1	8.2	45.2
PSPhSILi	58.4	5.7	26.2	9.7	–	48.8
PSDTTOLi	63.4	–	28.3	8.2	–	29.8

From Figures 2.2 – 2.4, it can be seen that all of these salts have a different ratio of carbon, nitrogen, oxygen, and sulfide, besides only PSTFSI-Li exhibits 8.2% F. However, none of the salts show up any elemental concentrations at lithium, that is because lithium ions are too tiny to be detected by the instrument. Thus, these results are rather consistent with the facts, which can be used as evidence to confirm the successful synthetic PSLSs.

2.3.1.3 Scanning Electron Microscopy. Based on both size and functional group (Figure. 2.5), TFSI and PhSI modified nanoparticles have the similar particles size at around 500nm. As well the DTTO modified nanoparticles are more potent in reducing R_m than the other two particles at around 350nm, which tend to aggregate more easily. These perfect nano-sized particles can assistance lithium ions shuttle back and forth between the polymers chains more straight forwardly [81].

It is also detected from the images that all of the PSLSs have significant numbers of randomly distributed spherical grains and the maximum number of pores with a tiny size of the order of 500 – 1000nm, which is responsible for the ionic conductivity. The increased number of porosity leads to entrapment of large volumes of the plasticizers and salts in the pores secretarial for the increased conductivity. Finally, the SEM photograph of the smooth surface of the samples also reveals that the synthesized PSLSs in this study have a real compatible nature.

2.3.2 Film processing

Poly (ethylene glycol 1000 dimethyl ether) was incorporated with PEO based lithium salt to form the film by the method of grinding. The nicely grinded mixture was then folded, and placed in between two Teflon coated sheets, then hot pressed in a Carver press at 100℃ under 10 M psi pressure [83]. Two thin stainless steel plates were used as a spacer to control the thickness of the film. The polymer films were cut circularly in an area of

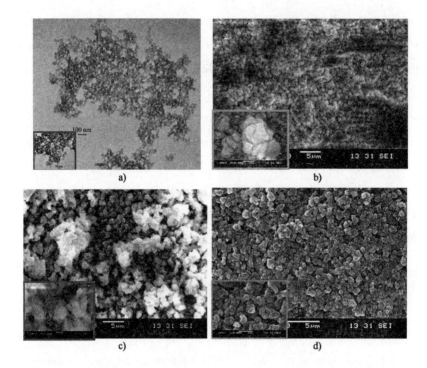

Figure 2.5. a) TEM for Polystyrene nanoparticles, the diameter of the particles is around 50nm; b) SEM for PSTFSILi nanoparticles, the diameter of the particles is around 500nm; c) SEM for PSPhSILi nanoparticles, the diameter of the particles is around 500nm; d) SEM for PSDT-TOLi nanoparticles, the diameter of the particles is around 350nm [82] [83].

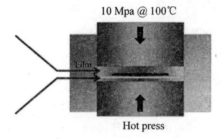

Figure 2.6. Schematic view of hot-pressing step of SPE Films

2.04 cm^2 and sandwiched between two steel electrodes and subjected to the impedance analyzer (Figure 2.6). The various weight ratios of PEO-based SPEs films along with PSLSs are respectively listed in Tables 2.2. −2.4.

Table 2.2. Formulations of the SPEs films with various weight ratios of PSTFSILi and LiTFSI[a]

Components weight (mg)				Abbreviation of the formulation
PSTFSILi	PEO	PEG	LiTFSI	
0	75	100	50	PPL = 3 : 4 : 2
25	75	100	25	TFSI-PPL = 1 : 3 : 4 : 1
25	75	100	50	TFSI-PPL = 1 : 3 : 4 : 2
25	75	100	75	TFSI-PPL = 1 : 3 : 4 : 3
25	37.5	50	12.5	TFSI-PPL = 2 : 3 : 4 : 1
25	37.5	50	25	TFSI-PPL = 2 : 3 : 4 : 2
25	37.5	50	37.5	TFSI-PPL = 2 : 3 : 4 : 3
25	25	37.5	8.3	TFSI-PPL = 3 : 3 : 4 : 1
25	25	37.5	16.7	TFSI-PPL = 3 : 3 : 4 : 2
25	25	37.5	25	TFSI-PPL = 3 : 3 : 4 : 3

a. During sampling process, bias (± 0.2mg) occurs.

Table 2.3 Formulations of the SPEs films with various weight ratios of PSPhSILi and LiTFSI

Components weight (mg)				Abbreviation of the formulation
PSPhSILi	PEO	PEG	LiTFSI	
25	75	100	25	PhSI-PPL = 1 : 3 : 4 : 1
25	75	100	50	PhSI-PPL = 1 : 3 : 4 : 2
25	75	100	75	PhSI-PPL = 1 : 3 : 4 : 3
25	37.5	50	12.5	PhSI-PPL = 2 : 3 : 4 : 1
25	37.5	50	25	PhSI-PPL = 2 : 3 : 4 : 2

continued

Components weight (mg)				Abbreviation of the formulation
PSPhSILi	PEO	PEG	LiTFSI	
25	37.5	50	37.5	PhSI-PPL = 2 : 3 : 4 : 3
25	25	37.5	8.3	PhSI-PPL = 3 : 3 : 4 : 1
25	25	37.5	16.7	PhSI-PPL = 3 : 3 : 4 : 2
25	25	37.5	25	PhSI-PPL = 3 : 3 : 4 : 3

Table 2.4 Formulations of the SPEs films with various weight ratios of PSDTTOLi and LiTFSI

Components weight (mg)				Abbreviation of the formulation
PSDTTOLi	PEO	PEG	LiTFSI	
25	75	100	25	DTTO-PPL = 1 : 3 : 4 : 1
25	75	100	50	DTTO-PPL = 1 : 3 : 4 : 2
25	75	100	75	DTTO-PPL = 1 : 3 : 4 : 3
25	37.5	50	12.5	DTTO-PPL = 2 : 3 : 4 : 1
25	37.5	50	25	DTTO-PPL = 2 : 3 : 4 : 2
25	37.5	50	37.5	DTTO-PPL = 2 : 3 : 4 : 3
25	25	37.5	8.3	DTTO-PPL = 3 : 3 : 4 : 1
25	25	37.5	16.7	DTTO-PPL = 3 : 3 : 4 : 2
25	25	37.5	25	DTTO-PPL = 3 : 3 : 4 : 3

PEO : PEG : LiTFSI (abbreviation as PPL) at the weight ratio of 3 : 4 : 2 was prepared as a standard membrane. The films of the SPEs maintain good ductility and mechanical strength and do not snap upon appreciable manual bending or stretching while increasing the ratio of LiTFSI becomes soft and sticky. The PPL complex at a weight ratio of 3 : 4 : 3 is rather hard to make a nice film, which intuitively proves that the incorporate cross-linked PSLS into the electrolytes is a right way to improve the film strength.

2.3.3 Thermogravimetric Analysis

Table 2.5 indicates that the three PSLSs are stable up to 400 ℃ and show good anti-thermal shrinkage performance. The TGA curves of PSLSs exhibit two mass loss steps. The initial mass loss below 400 ℃ is due to the gradual evaporation of absorbed moisture. The second mass loss from approximately 400 to 480℃ is the result of the decomposition of the salts. Our results show that the thermal stability PSPhSILi > PSDTTOLi > PSTFSILi, PSPhSILi shows the best thermal stability up to 450 ℃. In the case of PSTFSILi, due to the similar structure of LiTFSI, which may display more moisture sensitive, the total 11% weight loss before 400 ℃ is higher than the other two. PSPhSILi and PSDTTOLi have pretty similar thermal stability, below 400 ℃, there is no degradation. In a temperature range of 400 – 500℃ around 20% of the material is degraded, during this stage weight loss and volatilisation of degradation products take place rapidly. Polystyrene-based polymer chains may display random chain scission and then degrade into a mixture of styrene, toluene, methyl styrene, etc.

Table 2.5. Thermal degradation of PSLSs at different temperatures

Sample	Decomposition temperature, T_d (℃)	Percentage weight loss (%) at various temperature			
		200 ℃	300 ℃	400 ℃	500 ℃
PSTFSILi	376	4.5	6.6	11.1	29.9
PSPhSILi	450	0.7	2.1	2.7	22.5
PSDTTOLi	449	0.5	1.5	4.8	23.6

From Figure 2.7 it is observed that polystyrene degradation occurs in a single step, the degradation starts at about 450 ℃ and at 480 ℃ degradation is almost complete. The differences in degradation behavior in different PSLSs are also dependent on the functional group, the presence of aromatic benzene

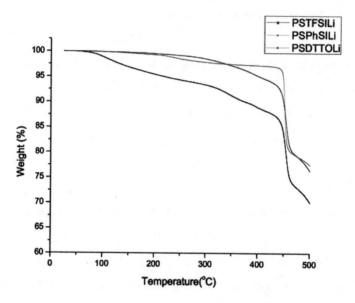

Figure 2.7. TGA curves under air flow for PSTFSILi, PSPhSILi & PSDTTOLi

ring enables PSPhSILi more resistant to thermal degradation compared to other functional groups.

2.3.4 Ionic Conductivities

As far as we know, Renaud Bouchet1 group reported a single-ion BAB triblock copolymer, P(STFSILi)-PEO-P(STFSILi) material, as a polymer electrolyte, exhibits 1.3×10^{-5} S/cm [84]. At the meantime, Shaowei Feng and co-workers developed the similar single lithium-ion conducting polymer electrolytes based on poly[(4-styrenesulfonyl)(trifluoromethane sulfonyl)imide] anions. However, the ionic conductivities could only reach 7.6×10^{-6} S/cm at 25 ℃ and 10^{-4} S/cm at 70 ℃ [85].

Nevertheless, in our research, the ionic conductivity of TFSI-PPL (Figure. 2.8.), at a weight ratio of 1 : 3 : 4 : 3, was 3.0×10^{-3} S/cm at 70 ℃ and much better than that of PhSI-PPL, which is 1.27×10^{-3} S/cm. And our best ionic conductivity of DTTO-PPL at a weight ratio of 1 : 3 : 4 : 3 is

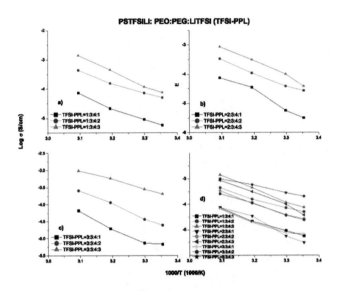

Figure 2.8. Ionic Conductivity versus Temperature of various weight ratios of PSTFSILi-PEO-PEG-LiTFSI (TFSI-PPL) composite membranes

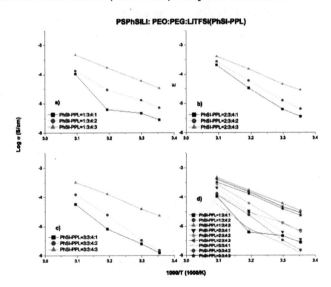

Figure 2.9. Ionic Conductivity versus Temperature of various weight ratios of PSPhSILi-PEO-PEG-LiTFSI (PhSI-PPL) composite membranes

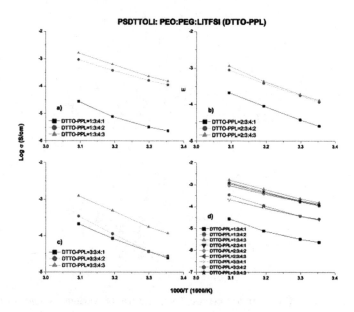

Figure 2.10. Ionic Conductivity versus Temperature of various weight ratios of PSPh-SILi-PEO-PEG-LiTFSI（DTTO-PPL） composite membranes

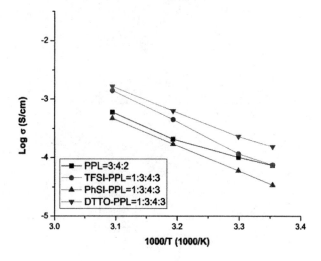

Figure 2.11. Ionic Conductivity versus Temperature of four best performance composite membranes

capable of reaching 1.6×10^{-3} S/cm at 50 ℃. The excellent performance of our electrolytes is rather competitive compared with those of other groups.

For these three designed PSLSs, although both structures of PSTFSI$^-$ and PSPhSI$^-$ provide additional nitrogen anionic sites for the cations, the higher ionic dissociation of the TFSI functional group than PhSI functional group, due to the presence of C-F bond which induces a strong electron-withdrawing effect, is coupled with the delocalization of the negative charge. According to the speculation, the existence of three strong electron withdrawing groups ($-SO_2$) may provide PSDTTO anions the best dissociating ability to enable the lithium ions moving.

Meanwhile, as the temperatures (Figures. 2.8, 2.9, & 2.10) are elevated, where the movement of the polymer chain segments rapidly increases, all the salts will follow the well established trend of increasing conductivity with the increased temperature. Nevertheless, with the increasing weight ratio of PSLSs, the conductivity clearly decreases. This trend is probably attributed to the high salt concentration with growing influence of the ion pairs, ion triplets, and the higher ion aggregations, which reduce the overall mobility and the number of effective charge carriers [2].

An additional favorable conduction pathway for the cations is created in the vicinity of the modified functional group, which may involve anionic sites on polystyrene surface as well as ether oxygens. Also, attributable to the cross-linked structure and chain extension, the cation transport is better assisted by the segmental motion, resulting in increased ion mobility and the bonding quality. Simultaneously, the enhanced composite membrane strengthens while ionic salts and plasticizers can also be contained in the polymer networks. So during the cationic transport process, the bonds between Li cations and anions in polymer electrolytes are also subjected to breaking and making more simply.

Furthermore, the incorporation of a plasticizer into a polymer matrix lowers the T_g of the polymer and helps the movement of the ions in the polymer, leading to an effective increase in ionic conductivity [86]. This T_g-conductivity relationship should be quite useful and helpful for the confirmation of experimental data, and is also effective to discuss the carrier ion concentration.

2.3.5 Differential Scanning Calorimetry

The DSC results are presented in Figure 2.12 and the essential glass transition and melting points for various electrolytes are listed in Table 2.6.

Table 2.6. Thermal properties of PSLSs based electrolytes membranes

Sample	a	b	c	d	e	f	g
T_g (℃)	-84.8	-78.2	-85.2	-85.3	-82.2	-85	-89.2
T_m (℃)	61.7	55.2	44.8	52.2	43.7	52.8	47.7

According to the DSC results of Figure 2.12. b & c, our electrolyte systems appear to behave better, suggesting that the contribution to the conductivity enhancement from the segmental flexibility of the PEO chains could be affected by increasing the more ratio of LiTFSI.

In Fig. 10. h, comparing PPL membrane with those of the addition of three different PSLSs, the values of T_g of these four membranes are decreased in the order of PPL (T_g = -84.8 ℃) > PhSI-PPL (T_g = -85 ℃) > TFSI-PPL (T_g = -85.2 ℃) > DTTO-PPL (T_g = -89.2 ℃); along with the values of T_m decreased in the order of PPL (T_m = 61.7 ℃) > PhSI-PPL (T_m = 52.8 ℃) > DTTO-PPL (T_m = 47.7 ℃) > TFSI-PPL (T_m = 44.8 ℃). It is fascinating to note that the decrease of T_g and T_m with adding the PSLSs in the electrolytes, among all, PSDTTOLi gives the best performance. The

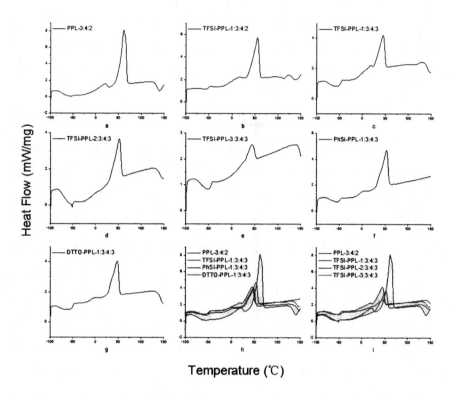

Figure 2.12. Heat flow versus Temperature of DSC curve: a) weight ratio of PEO : PEG : LiTFSI (PPL) = 3 : 4 : 2; b) weight ratio of PSTFSILi: PEO : PEG : LiTFSI (TFSI-PPL) = 1 : 3 : 4 : 2; c) weight ratio of TFSI-PPL = 1 : 3 : 4 : 3; d) weight ratio of TFSI-PPL = 2 : 3 : 4 : 3; e) weight ratio of TFSI-PPL = 3 : 3 : 4 : 3; f) weight ratio of PSPhSILi: PEO : PEG : LiTFSI (PhSI-PPL) = 1 : 3 : 4 : 3; g) weight ratio of PSDTTOLi: PEO : PEG : LiTFSI (DT-TO-PPL) = 1 : 3 : 4 : 3; h) four best performance complex membranes; i) various weight ratio of PSTFSILi

result of the addition of PSLSs to the electrolyte composition leads us to conclude that the guest PSLSs salt exerts a plasticizing effect on the polymer structure in this system, which increases the free volume available to polymer chain segments and therefore allows greater internal chain rotation and an increase in the segment flexibility and mobility [87]. The large, flexible anions of cross-linked PSLSs are believed to act as a plasticizer inhibiting recrystallization kinetics and thereby retaining the amorphous phase down to ambient temperatures. [29]

A further increase in salt content beyond the salt composition associated with the electrolyte with lowest T_g might result in an increase in polymer rigidity, where the interactions between the cations and polar segments belonging to different polymer chains reason to ionic crosslinking which hinders segmental mobility and shifts the T_g of the electrolyte to much higher temperatures. That is in contract with the behavior detected in TFSI-PPL electrolyte systems (include Figure 2.12.a, c, d, e in Figure 2.12.i). With the increasing weight ratio of PSTFSILi, the T_g of the membranes lowers down from -85.2 ℃ to the lowest degree -85.3 ℃, and soon increases to -82.2 ℃.

When came to T_m (Figure 2.12.h), the addition of PSLSs caused a change in the shape of the endothermic peak, and the peaks slightly shifted towards lower temperature. These observations clearly suggest that a significant contribution to conductivity enhancement comes from structural modifications associated with the polymer host caused by the salts. The reorganization of polymer chain may be hindered by the cross-linking centers formed by the interaction of the functional groups.

These figures also show a quite good relation in which the ionic conductivity increases with lowering the T_g for the systems having sufficient carrier ion concentration. The T_g of the polymer is directly related to the flexibility of the polymer chains, which also affects the ionic conductivity of the electrolyte. The lower the T_g, the higher the flexibility and conductivity. While incorporation of the electrolyte salts, the addition of plasticizer effectively compensates by playing a significant role in lowering the T_g of the polymer by enhancing the mobility of its chain segments. With amorphous domain in polymer electrolytes where ion conduction occurs and ion dissociation of the polymer electrolyte increases, the ionic conductivity is highly increased. This trend is prominent in our studies.

2.3.6　Cyclic Voltammetry

The characteristic voltammetry curves for each of these three best performance membranes are clearly shown in Figure 2.13. According to the CV measurements, both the films of PhSI-PPL & DTTO-PPL at the weight ratio of 1 : 3 : 4 : 3 exhibit high voltage stability up to 4.2 V. PhSI-PPL membrane shows an anodic peak at 1.26 V and, along with DTTO-PPL membrane exhibiting another anodic peak at 1.31 V [88]. The exciting results are probably credited to the dopant anions of the PSLSs [89], because the larger anions are more strongly bound to the polymer and are less prone to be replaced by smaller ions. Thus, the cation movement becomes more dominant than anion movement. Meanwhile, the reduction peak currents are higher than those of the oxidation current, indicating that electrolyte ions move in the PhSI-PPL and DTTO-PPL film during reduction process and enhance the doping level [90]. In this case, it is found that with the largest dopant ions, only cations move in and out of the polymer chain during cycling.

What is more, TFSI-PPL at a weight ratio of 1 : 3 : 4 : 3 displays the best electrochemical stability, as no significant electrochemical activity is shown up at potentials as high as 4.8 V versus Li/ Li$^+$. The rationale for designing PSTFSILi is the striking similarity with LiTFSI (as illustrated in Figure 2.14.) in the aspect that the negative charge on the nitrogen atom is delocalized by two sulfone groups and one trifluoromethane group, making less interionic attractions between cation and anion, commonly described as a low lattice energy salt [91]. Thus, it enables the electrolyte membrane shows

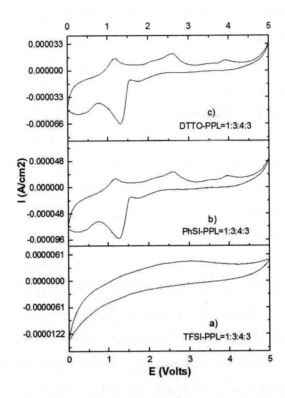

Figure 2.13. CV curves under Argon flow for electrolyte membranes: a) TFSI-PPL = 1 : 3 : 4 : 3; b) PhSI-PPL = 1 : 3 : 4 : 3; c) DTTO-PPL = 1 : 3 : 4 : 3

superior electrochemical stability.

Therefore, this TFSI-PPL solid polymer electrolyte also makes it potential to consider the use of high-potential cathode materials that cannot be safely tested in liquid electrolytes due to their lower electrochemical stability.

The stability of the PSLSs based hybrid electrolytes to achieve high voltages against lithium is confirmed, which ensures their capability for further use in high potential electrolyte materials.

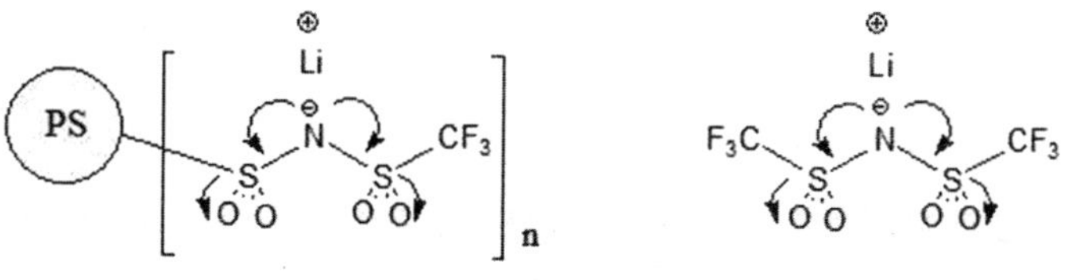

PSTFSILi   **LiTFSI**

Figure 2.14. Structural comparisons of PSTFSILi with LiTFSI

2.4 Summary

In this chapter, the synthesis of three different polystyrene lithium salts based SPEs has been accomplished for application in lithium ion batteries. These PSLSs based polymers show good ionic conductivity and excellent thermal stability. The polymers were made from relatively cheap starting materials.

Studies have shown that the obtained PSLSs based polymer films possess good ionic conductivity at ambient temperature, including excellent thermal and electrochemical stability, and mechanical stability. The highest ionic conductivity was able to reach with DTTO-PPL membrane at a weight ratio of 1 : 3 : 4 : 3 (1.54×10^{-04} S/cm at room temperature and 1.6×10^{-03} S/cm at 50 ℃). This performance is better than that of the Renaud Bouchet group's single-ion BAB triblock copolymers [84] and also more competitive with Shaowei Feng and co-workers' single lithium-ion conducting polymer electrolytes [85]. These PSLSs are considered to be useful for application to electrolyte membranes in battery devices. Further study of the applications is currently in progress.

Chapter 3 New Plasticized Low Lattice Energy Lithium Salts

3.1 Introduction

Nowadays, there is a compelling need for strategic design and development of an advanced solvent free electrolyte system, which is free of leakage and possesses high ionic conductivity and desired electrochemical and mechanical properties.

The availability of lithium rechargeable batteries featuring solvent free highly conductive solid electrolyte (SE) systems will have a major impact on the EV/HEV industry, leading to a significant reduction in environmental pollution and improved performance, compared with current carbonate based liquid electrolytes. The driving force for this advance is the strong potential for achieving high energy densities, high cell voltage, and superior self-discharge characteristics, while largely mitigating deficiencies, such as leakage, instability.

In the solid electrolyte (SE) systems, the Ionic conductivity of an electrolyte depends directly on the concentration of the lithium ion. Nevertheless, the increase in lithium salt (LS) concentration may not be helpful in improving the ionic conductivity, since the viscosity also plays an important role. Higher salt concentrations will lead to a viscous solution with lower ionic conductivity. The improvement in ionic conductivity with low salt concentrations can be accomplished by using salts with more than one lithium ion in their structures which is named as dilithium salts (DLSs).

With this vision, we have developed a facile synthetic strategy to prepare a new class of DLSs from relatively inexpensive starting materials. Owing to the presence of two lithium ions per molecule, these salts involve lower concentrations than normally used salts to achieve comparable ionic conductivities at ambient temperatures, which could also lower down the cost of batteries. The interesting feature of this series of salts is their striking similarity with LiTFSI, in the aspect that both sides are imides and the negative charges on the nitrogen atom are delocalized by two sulfone groups, making lithium ion highly mobile [92]. Hence, like LiTFSI, these salts are expected to display low inter-ionic attractions between cation and anion (i. e., low lattice energy), leading to high ionic conductivity. On the other hand, the ethylene oxide (EO) unit is the best solvating medium. EO is made by C – O, C – C, C – H bonds, which are highly chemical, electrochemical and mechanically stable. Moreover, the repeated unit – CH_2CH_2O – provides a suitable space for Li-ion to make chelating with oxygen, and Li-ion migration is associated with the segmental mobility of EO chains [93].

3.2 Experimental

3.2.1 Materials

Diethylene glycol, triethylene glycol, tetraethylene glycol, pentaethylene glycol, 1, 4 - dioxane, thionyl chloride, pyridine, thiourea, trifluoromethane sulfonimide, lithium hydroxide monohydrate anhydrous, triethyl orthoformate, 1, 2 - ethanedithiol, ethanol, hydrogen peroxide, lithium methoxide, and acetic acid were purchased from Sigma-Aldrich Chemical Co., Ltd (USA). Lithium bis (trifluoromethyl sulfonyl) imide (LiTFSI) was obtained from 3M. Acetone, acetonitrile, dichloromethane, methanol and the other solvents were supplied by Fisher Scientific Co., Ltd., (USA).

3.2.2 General procedures for the preparation of LS-1

Lithium (oxy bis (ethane - 2, 1 - diylsulfonyl)) bis (((trifluoromethyl) sulfonyl) amide): LS - 1

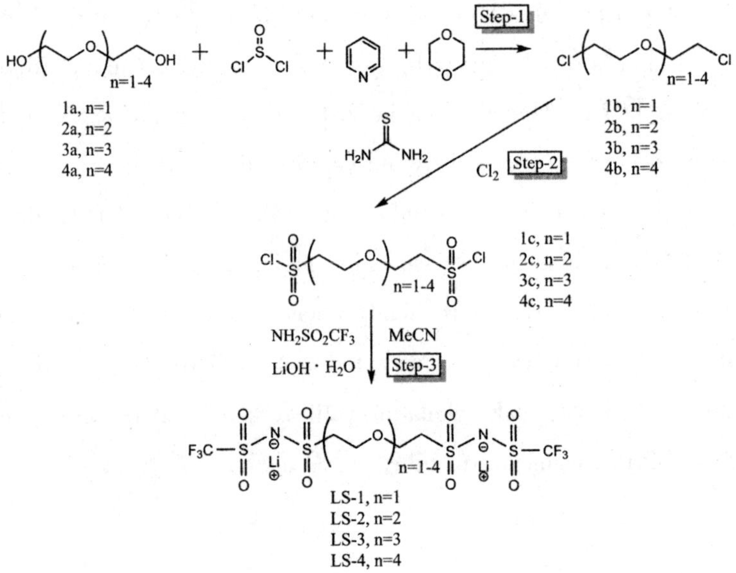

Scheme 3.1. Synthetic Routes of dilithium salts (LS-1~4)

3.2.2.1 Synthesis of bis (chloroethyl) ether (1b). Diethylene glycol (1a, 2.41 g, 22.7 mmol) was dissolved in 1, 4 - dioxane (2.8 ml, 32.83

mmol). Thionyl chloride (3.95 ml, 54.42 mmol) was slowly added together with two drops of pyridine to it. The mixture was stirred at 120 ℃ for 12 hours and then cooled down to room temperature. After filtration, the mixture was distilled at 100 ℃ under reduced pressure. The residue was placed under a high vacuum at 70 ℃ overnight to give the desired compound bis (chloroethyl) ether (1b, yield: 1.80 g, 55.4%) as a yellow liquid.

3.2.2.2 Synthesis of 2, 2' - oxy bis (ethane - 1 - sulfonyl chloride). Bis (chloroethyl) ether (1b, 1.78 g, 12.6 mmol), thiourea (1.92 g, 25.2 mmol) and 8 ml of ethyl alcohol were added in a 50 ml flask. The mixture was heated to 90 ℃ for reflux overnight. The reflux reaction was cooled to room temperature, and transfered to a 100 ml three neck flask. 1000 parts of water and 500 parts of cracked ice (35 ml) were added to it. In another setup, adding dropwise concentrated hydrochloric acid in a small quantity of $KMnO_4$ (4.10 g) generates chlorine. Then the chlorine was passed into the resulting solution, while maintaining the temperature below 10 ℃. The solution was distilled under reduced pressure at 80 ℃. 2, 2' - Oxy bis (ethane - 1 - sulfonyl chloride) (1c, yield: 3.06 g, 89.7%) was obtained in the form of a heavy yellow oil;

3.2.2.3 Synthesis of LS - 1. Lithium (oxy bis (ethane - 2, 1 - diyl-sulfonyl)) bis (((trifluoromethyl) sulfonyl) amide): LS - 1.

2, 2' - Oxy bis (ethane - 1 - sulfonyl chloride) (1c, 3.06 g, 11.3 mmol), trifluoromethanesulfonamide (3.53 g, 23.7 mmol), lithium hydroxide monohydrate anhydrous (1.94 g, 46.2 mmol) were placed in a 50ml round-bottomed flask, and 25 ml of anhydrous acetonitrile was added to it. The mixture was stirred at 100 ℃ overnight. The solution was filtered, and the filtrate was concentrated with a rotary evaporator. The resultant solid was dissolved in dichloromethane (30 mL). The precipitated salt was filtered off, and the filtrate was concentrated again. The residue was placed under high vacuum at 70 ℃ overnight to give the final compound as a white solid (LS – 1, yield: 4.66 g, 81.2%)

IR: 2204.67 cm^{-1} (w), 1664.26 cm^{-1} (S-N-S, m), 1473.81 – 1412.02 cm^{-1} (C – H, v), 1268.76 – 1193.0 cm^{-1} (S = O, s), 1103.48 cm^{-1} (C – O, s), 988.46 cm^{-1} (C – F, s), 624.1 cm^{-1} (S-N, s).

3.2.3 General procedures for the preparation of LS – 2

Lithium ((ethane – 1, 2 – diylbis (oxy)) bis (ethane – 2, 1 – diylsulfonyl)) bis (((trifluoromethyl) sulfonyl) amide): LS – 2. An analogous procedure was used to prepare 2b started with triethylene glycol (2a, 2.42 g, 16.1 mmol), 1, 4 – dioxane (2.7 ml, 31.65 mmol), thionyl chloride (2.8 ml, 38.6 mmol), except that the filtration was distilled at 120 ℃ to give the desired compound as a yellow liquid (2b, Yield: 2.64 g, 87.6%).

The similar method was used to synthesise the second step 2c through (2b, 2.64 g, 14.1 mmol), thiourea (2.15 g, 28.2 mmol) and 8ml of ethyl alcohol, except that the filtration was distilled at 100 ℃ to give the desired compound as a yellow oil (2c, Yield: 3.93 g, 88.3%).

The corresponding lithium salt was prepared analogously from 2c (3.93 g, 12.5 mmol), trifluoromethanesulfonamide (3.9 g, 26.2 mmol), lithium hydroxide monohydrate anhydrous (2.15 g, 51.2 mmol) to give the desired compound as a white solid (LS – 2, yield: 5.73 g, 83.2%);

IR: 2206.86 cm^{-1} (w), 1654.85 cm^{-1} (S – N – S, m), 1473.80 – 1405.99 cm^{-1} (C – H, v), 1271.84 – 1193.27 cm^{-1} (S = O, s), 1103.48 cm^{-1} (C – O, s), 988.55 cm^{-1} (C – F, s), 622.78 cm^{-1} (S – N, s).

3.2.4 General procedures for the preparation of LS-3

Lithium (((oxy bis (ethane-2, 1-diyl))) bis (oxy))) bis (ethane -2, 1-diylsulfonyl)) bis (((trifluoromethyl) sulfonyl) amide): LS-3. An analogous procedure was used to prepare 3b started with tetraethylene glycol (3a, 2.41 g, 12.4 mmol), 1, 4-dioxane (2.8 ml, 32.83 mmol), thionyl chloride (3.95 ml, 54.42 mmol), except that the filtration was distilled at 120 ℃ to give the desired compound as a yellow liquid (3b, yield: 2.31 g, 80.6%);

The similar method was used to synthesise the second step 3c through (3b, 2.31 g, 10.0 mmol), thiourea (1.52 g, 20.0 mmol) and 8ml of ethyl alcohol, except that the filtration was distilled at 100 ℃ to give the desired compound as a yellow oil (3c, yield: 2.50 g, 70%);

The corresponding lithium salt was prepared analogously from 3c (2.50 g, 7.0 mmol) and trifluoromethanesulfonamide (2.08 g, 13.9 mmol), lithium hydroxide monohydrate anhydrous (1.20 g, 28.6 mmol) to give the desired compound as a pale yellow solid (LS-3, yield: 3.68 g, 88.7%).

IR: 2206.86 cm^{-1} (w), 1664.68 cm^{-1} (S-N-S, m), 1473.84 - 1375.33 cm^{-1} (C-H, v), 1255.28 - 1191.02 cm^{-1} (S=O, s), 1101.93 cm^{-1} (C-O, s), 990.75 cm^{-1} (C-F, s), 625.80 cm^{-1} (S-N, s).

3.2.5 General procedures for the preparation of LS-4

Lithium 3, 6, 9, 12-tetraoxatetradecanedisulfonylbis (((trifluoromethyl) sulfonyl) amide): LS-4. An analogous procedure was used to prepare 4b started with tetraethylene glycol (4a, 2.39 g, 10.1 mmol), 1, 4-dioxane (2.7 ml, 31.65 mmol), thionyl chloride (1.8 ml, 24.78 mmol), except that the filtration was distilled at 120 ℃ to give the desired compound as a yellow liquid (3b, yield: 2.24 g, 81.2%).

The similar method was used to synthesise the second step 4c through (4b, 2.24 g, 8.13 mmol), thiourea (1.24 g, 16.3 mmol) and 8ml of ethyl alcohol, except that the filtration was distilled at 100℃ to give the desired compound as a yellow oil (4c, yield: 2.81 g, 85.6%).

The corresponding lithium salt was prepared analogously from 4c (2.81 g, 7 mmol) and trifluoromethanesulfonamide (2.07 g, 13.9 mmol), lithium hydroxide monohydrate anhydrous (1.20 g, 28.6 mmol) to give the desired compound as a heavy white solid (LS-4, yield: 3.87 g, 86.7%).

IR: 2207.80 cm^{-1} (w), 1637.65 cm^{-1} (S-N-S, m), 1473.68 – 1405.23 cm^{-1} (C-H, v), 1254.57 – 1187 cm^{-1} (S=O, s), 1102.22 cm^{-1} (C-O, s), 1007.40 cm^{-1} (C-F, s), 653.77 – 530.90 cm^{-1} (S-N, s).

3.3 Results and Discussion

3.3.1 Film process

The Poly (ethylene glycol 1000 dimethyl ether) was incorporated with PEO based lithium salt to form the film by the method of grinding. The nicely ground mixture was then folded, and placed in between two Teflon coated sheets, then hot pressed in a Carver press at 100℃ under 5000 psi pressure [83]. Two thin stainless steel plates were used as a spacer to control the thickness of the film. The polymer films were cut circularly in an area of 2.04 cm^2 and sandwiched between two steel electrodes and subjected to the impedance analyzer (Figure 2.6). The various weight ratios of PEO-based SEs films along with LSs are listed in Tables 3.1. – 3.2.

3.3.2 Ionic Conductivities

For these four designed DLSs, all of their structures providing additional nitrogen anionic sites for the cations, due to the presence of C-F bond of the TFSI functional group which induces a strong electron-withdrawing effect, coupled with the delocalization of the negative charge, providing them the high ionic dissociation.

Table 3.1 Formulations of the SEs films with various weight ratios of LS-1 and LiTFSI[a]

Components weight (mg)				Abbreviation of
LS-1	PEO	PEG	LiTFSI	the formulation
25	75	100	25	SE1
25	75	100	50	SE2
25	75	100	75	SE3
25	37.5	50	12.5	SE4
25	37.5	50	25	SE5
25	37.5	50	37.5	SE6
25	25	37.5	8.3	SE7

a. During sampling process, bias (± 0.2mg) occurs.

Table 3.2 Formulations of the SEs films with various weight ratios of LSs and LiTFSI

Components weight (mg)						Abbreviation of the
LS-1	LS-2	LS-3	LS-4	PEO	PEG	formulation
75	-	-	-	75	100	SE8
-	75	-	-	75	100	SE9
-	-	75	-	75	100	SE10
-	-	-	75	75	100	SE11

In our research, the performance of these DLSs-based electrolytes is pretty good, and they show rather high ionic conductivity. As the temperatures (Figures 3.1 & 3.2) are elevated, where the movement of the segments in the membrane rapidly increases, all the salts follow the well established trend of increasing conductivity with the increased temperature. Meanwhile, with the increasing weight ratio of LiTFSI in SE – 1, SE – 2, SE – 3, the conductivity clearly increases and with the increasing weight ratio of LS – 1 in SE – 1, SE – 4, SE – 7, the conductivity also increases. This trend is probably attributed to the high salt concentration with growing influence of the ion pairs, ion triplets, and the higher ion aggregations, while more number of effective charge carriers is provided [2].

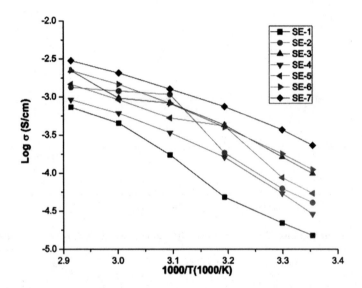

Figure 3.1. Ionic Conductivity versus Temperature of various LS – 1 based SE membranes

While comparing the four LSs based SE membranes (Figure 3.2), SE – 10 with three EO chains in the structure shows the best ionic conductivity, which is able to reach 1.09×10^{-04} S/cm at room temperature and 2.51×10^{-03} S/cm at 70 ℃. Like the similar structure of LiTFSI, these salts are expected to display low inter-ionic attractions between cation and anion (i. e., low lattice energy), leading to a high ionic conductivity. The additional favora-

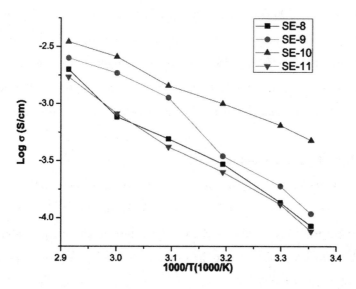

Figure 3.2. Ionic Conductivity versus Temperature of various LSs based SE membranes

ble conduction pathway for the cations is created in the vicinity of the modified functional group, which may involve anionic sites. On the other hand, attributable to the EO chain structure and chain extension, the cation transport is better assisted by the segmental motion, resulting in increased ion mobility and the bonding quality. Hence, EO chains are made by C – O, C – C, C – H bonds, which are highly chemical, electrochemical and mechanically stable. Moreover, the repeated unit – CH_2CH_2O – provides a suitable space for Li-ion to make chelating with oxygen, and Li-ion migration is associated with the segmental mobility of EO chains [93]. Simultaneously, the enhanced composite membrane strengthens while ionic salts and plasticizers can also be contained in the polymer networks. So during the cationic transport process, the bonds between Li cations and anions in DLSs based electrolytes are also subjected to breaking and making more simply.

Furthermore, the incorporation of a plasticizer into the solid electrolytes lowers the T_g of the membrane and helps the movement of the ions in the system, leading to an effective increase in ionic conductivity [86]. This T_g-conductivity relationship should be quite useful and helpful for the confirmation of experimental data, and is also effective to discuss the carrier ion concentration.

3.3.3 Cyclic Voltammetry

The characteristic voltammetry curves for each of the four LSs-based SE membranes are clearly shown in Figure 3.3. According to the CV measurements, both the films of SE – 9 & SE – 10 exhibit high voltage stability from – 2.0 V up to 4 V; SE8 exhibits in a range of – 1 ~ 3.5 V, SE11 performs in a range of – 1.5 ~ 4.0 V.

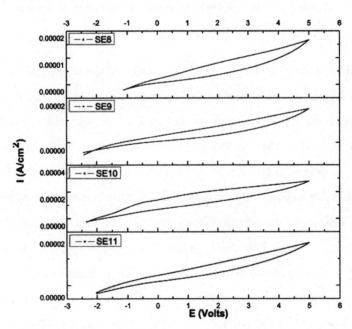

Figure 3.3. CV curves for four LSs-based SE membranes

What is more, SE10 displays the best electrochemical stability, as no significant electrochemical activity is shown up at potentials as high as 4.0 V versus Li/ Li$^+$. The rationale for designing of LS-3 is the striking similarity with LiTFSI in the aspect that the negative charge on the nitrogen atom is delocalized by two sulfone groups and one trifluoromethane group, making less interionic attractions between cation and anion, commonly described as a low lattice energy salt [91]. Thus, it enables the electrolyte membrane shows superior electrochemical stability. The exciting results are also probably credited to the dopant anions of the TFSI structure [89] because the larger anions are more strongly bound to the PEO and PEG system and are less prone to be replaced by smaller ions. Thus, the cation movement becomes more dominant than anion movement. Meanwhile, the reduction peak currents are higher than those of the oxidation current, indicating that electrolyte ions move in the SE film during reduction process and enhance the doping level [90]. In this case, it is found that only cations move in and out of the polymer chain during cycling.

Therefore, the stability of the DLSs based electrolytes to achieve high voltages against lithium is confirmed, which ensures their capability for further use in high potential electrolyte materials that cannot be safely tested in liquid electrolytes due to their lower electrochemical stability.

3.4 Summary

In this chapter, the synthesis of fluorinated dilithium salts has been accomplished for application in lithium ion batteries. These fluorinated dilithium salts show good ionic conductivity, excellent thermal stability and electrochemically stable below 4.2V versus lithium metal. The salts were made from relatively cheap starting materials.

Studies have shown that the obtained dilithium salts based films show better ionic conductivity than PEO and PEG based electrolytes at ambient temperature. Our electrolytes also show excellent thermal and electrochemical stability, as well as mechanical stability. The highest ionic conductivity is able to reach with the SE10 membrane (1.09 x10^{-04} S/cm at room temperature and 2.51 x10^{-03} S/cm at 70 ℃). These fluorinated dilithium salts are considered to be useful for application to electrolyte membranes in battery devices. Other characterization studies are future endeavors to be accomplished, and further study of the applications is currently in progress.

Chapter 4 New Nonflammable Di-cationic Ionic Liquids

4.1 Introduction

Over the last several years, since the discovery of air-and water-stable ionic liquids (ILs) by Wilkes in 1992 [49], which have been widely promoted as more reliable class of "green solvents" compared with organic carbonates, attribute to their unique properties, such as thermal stability, high electrical conductivity, high polarity [50], and negligible vapor pressures [51], coupled with a wide liquid range, ILs are recently fascinating considerable attention for applications in numerous fields, though they were described almost a century ago.

ILs are a class of designable organic compounds which displays two characteristic structures: firstly, they consist only of ions which are poorly coordinated. Secondly, they are liquids below 100 ℃ [52], if the ILs are liquid at room temperature, they are called room temperature ionic liquids (RTILs) [53]. Through combining organic cations with suitable anions allows to tailoring the appropriate physical, chemical and biological properties for ILs, while some even possess unexpected functions resulting from synergetic collaboration between the two components. Hence, the attractive flexibility or 'tunability' in the design has driven phenomenal interest in ILs synthesis [54]. The unique structure and performance of ILs as a platform not only offers additional breaks to modify these ionic materials' physical properties (e. g. melting point, density, polarity, viscosity, hydrophobicity/hydrophilicity, solubility) for specific applications, but also offers other gorgeous chemical features such as fundamental ionic conductivity, tremendous thermal, chemical, and electrochemical stability [55].

By either a physical combination of ILs or chemical modification (covalent functionalization or ion-exchange metathesis process of the ionic constituents, specific functional groups can be easily incorporated into the ILs skeleton), a number of IL-containing composite materials and functional ILs have been effectively realized and applied to the enormous area.

First of all, nowadays notable efforts for ILs have been made focusing on the design of safer and more environmental kindly solvents [56], for conventional organic solvents are often toxic, flammable and volatile. Compared with the volatile organic compounds, ILs display excellent dissolution performance for organic, inorganic, and polymer materials, and their immeasurable vapor pressure [57] and non-flammability properties provide them the ability to avoid atmospheric pollution because there would be no loss of solvent through evaporation. The thermal decomposition temperatures higher than 300 ℃ could enhance their recycling efficiency [58]. Accordingly, ILs can be applied to replacing traditional volatile organic solvents for a host of practices linked to green chemistry and clean technology such as organic reactions [59], extraction, catalysis, and separation processes [55].

Secondly, ILs have a very broad range of viscosities, may vary between 20 and 40,000 cp compared with viscosities of typical organic solvents which are in a range of 0.2 and 100 cp. In ILs, only minor change of structure may result in a significant change in viscosity. Thus, ILs are discovered extensive use as engineering fluids or as innovative lubricating systems [50]. IL-based materials could straightforwardly be operated under extreme conditions such as high or low temperature and high vacuum or pressure, owing to their significant thermal stability in a quite wide liquid phase temperature range from 200 to 300 ℃, combined with designable excellent mechanical, chemical, and electrochemical stability.

Thirdly, the special designable structures and functional properties not only retain the key features of the original materials but also possess the characteristics of ILs. It is thought that the ionic nature of ILs, which the molecular materials lack, can offer these materials with intrinsic ion conductivity as well as a substantial ionic skeleton for producing progressive materials. At the meantime, ILs exhibit perfect electrochemical stability in a wide electrochemical window of 2 ~ 5 V [60]. Thus, ILs, especially RTILs, would be excellent candidates for prospective applications as encouraging electrolyte bases or additives in lithium secondary batteries and other energy-related applications, such as fuel cells, supercapacitors [61], and dye-sensitized solar cells [62] and [63].

Since many factors can affect their conductivity, such as viscosity, density, ion size, anionic charge delocalization, aggregations and ionic motions, ILs are expected as key materials which might give a solution to the safety problems of batteries due to their non-flammable property. However, the selectivity of carrier ion transport is one of the problems when ILs are applied to batteries. Since ILs do not include electroactive species, it is necessary to add salts or acids before their use. The mobility of electroactive ions has great effects on the electrical power of the batteries as lithium cation does on lithium-ion batteries, while the introduction of ethylene oxide (EO) chains attached to the methyl pyrrolidine, butyl-imidazolium ring or diethyl-sulfide may lead to lower glass transition temperature, and also improve ionic conductivity [41].

As a new family of ILs, DILs are consist of the anion and the doubly charged cation which is composed of two singly charged cations as head groups linked by a variable length of alkyl or oligo ethylene glycol chain as a rigid or flexible spacer. Thus, the DILs matrix created offers the opportunity to investigate (a) the influence of cation and anion variation, and (b) the influence of the chain length.

Recently, most of there searchers have focused on mono cationic-type ILs, although the number of DILs described in the literature is increasing [62]. Anderson et al. presented the synthesis and characterization of a variety of DILs containing alkane spacers [65]. Kubisa and Biedron investigated DILs obtained by the functionalization of PEG with triphenylphosphine [66]. Ohno and co-workers have shown that dicationic PEG-based molten salts are bearing imidazolium cations as ion conducting materials [67]. Dicationic and polycationic ILs linked by alkyl chains display extraordinarily higher chemical and thermal stability than their mono-cationic analogs published by Armstrong and coworkers [65]. It has also been presented that the acute toxicity of DILs is in several cases below the levels detected for those mono-cationic and that the use of head groups connected via polyethylene glycol could be identified as structural elements reducing the toxicity [50].

Based on the literature, DILs possess distinctive features in critical micelle concentration [57], such as a wider liquid range and higher thermal stability, better behavior as electrolytes characteristics, and higher thermal stabilities than mono-cationic ILs and other traditional solvents [68]. As we know, ILs should have the capability to dissolve more Li salt to have a higher Li^+ conductivity, if ILs are intended to be applied to Li batteries. Thus, by incorporating a PEO oligomer into the DILs structure, the conductivity could be enhanced primarily by improving cation transport in the PEO segment, although the viscosity of DILs does not decrease [69] due to the weak interaction of ethylene oxide segment compared to the alkyl chain [70].

For DILs, the relationship studies between their structure and physicochemical characteristics and molecular structures are, as yet, still relatively rare [71]. Therefore, it is urgent and necessary to explore other new DILs structures to gain further understanding and extend the applications of DILs as electrolyte components.

In this paper, we involved the strategic design and synthesis of TFSI-based & DCA-based RTILs to address the PSS problem. The solvent-in-salt (SIS) electrolytes display outstanding suppression of lithium polysulfide (Li_2S_n) dissolution via the use of heavily saturated electrolyte solutions; however, the common electrolyte solvents (1:1 DME/DOL) used are extremely flammable [94]. By contrast, TFSI based DILs have attracted great attentions due to their non-flammable & board thermal stability, although in general TFSI based ILs cannot dissolve enough lithium salt (typically <0.5M) to form the relatively high concentration. We place the hope to improve the solubility of lithium salt, through the inserting different length of EO chains into the two cationic moieties. Different from TFSI-based DILs, using DCA as the counter anion for the DILs, namely DCA-based DILs have much lower viscosity and tend to display much better ionic conductivity [95]. DCA anion is normally unusable for Li-ion battery application (4V electrochemical window), but can perhaps be used for Li-S batteries as a result of the much lower operating voltage (~3V) [96].

On the other hand, imidazolium cation based DILs are promising candidates to replace volatile solvents due to their chemical and thermal stability, non-volatility, high ionic conductivity, large electrochemical window and good solvent behavior [97]. Pyrrolidinium-based ILs show a better compatibility toward the lithium anode [98]. Lots of publications have been devoted to imidazolium-based, and pyrrolidinium-based DILs and the other DILs such as sulfonium-based DILs have seldom been discussed. Sulfonium cation-based ILs could show higher cell efficiencies because of lower charge transfer resistance [99] and [100]. Additionally, the EO units in the cation of the ionic liquid effectively provide relatively ion mobility which may contribute to high conductivity. The TFSI and DCA anions can facilitate stable charge/discharge of the Li-S cells.

The series of compounds can be a possible option to be used as co-solvents to improve further lithium salts solubility and the conductivity of the electrolyte which is necessary for capacity and rate performance of the Li rechargeable battery.

4.2 Experimental

4.2.1 Materials

Diethylene glycol, triethylene glycol, tetraethylene glycol, pentaethylene glycol, 1,4 – dioxane, thionyl chloride, pyridine, thiourea, 1 – methylpyrrolidine, 1 – butyl imidazolium, diethyl sulfide, potassium hydroxide, 4 – toluenesulfonyl chloride, lithium bis (trifluoromethyl sulfonyl) imide, sodium dicyanamide and Dowex © 22 were purchased from Sigma-Aldrich Chemical Co., Ltd (USA). All the solvents were supplied by Fisher

Scientific Co., Ltd., (USA).

4.2.2 General procedures for the preparation of MPy-based DILs

As shown in Scheme 4.1, the synthesis of MPy-based DILs involves three steps.

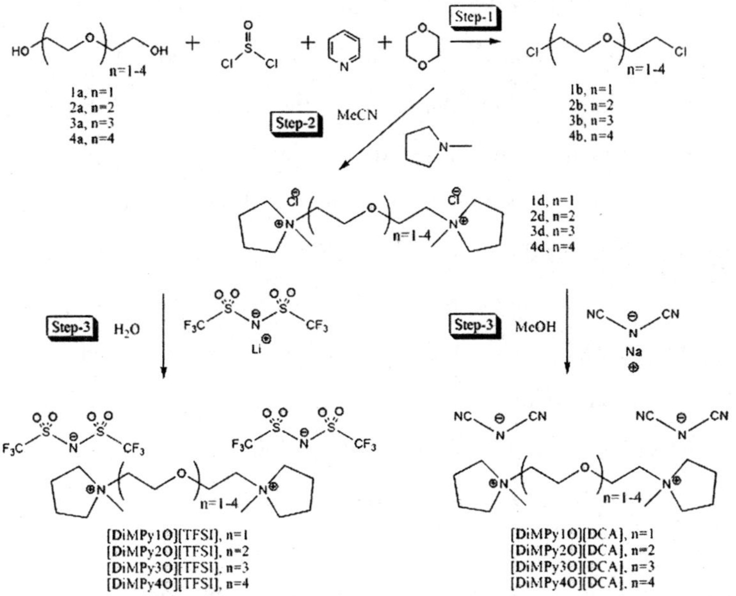

Scheme 4.1. Synthetic Routes of MPy-based DILs

4.2.2.1 Synthesis of DILs [DiMPy1O] [TFSI]. 1, 1'- (oxy bis (ethane-2, 1-diyl)) bis (1-methylpyrrolidin-1-ium) bis ((trifluoromethyl) sulfonyl) amide: [DiMPy1O] [TFSI]. Bis (chloroethyl) ether (1b, 4.06 g, 28.4 mmol) and 1-methylpyrrolidine (5.13 g, 60.2 mmol) were dissolved in 10 ml dried acetonitrile and refluxed for 4h under stirring. The mixture was cooled down to room temperature and stirred overnight. After the solvent had been removed by vacuum evaporation, the residue was washed three times with 15 ml of ethyl ether. The remaining volatile was removed under high vacuum to give the desired compound 1, 1'- (oxy bis (ethane −2, 1 −

diyl)) bis (1 - methylpyrrolidin - 1 - ium) dichloride (1d, yield: 8.05 g, 90.5%) as a brown liquid.

Halide anion exchange of DILs with [TFSI]$^-$ and [DCA]$^-$ anions, was respectively performed according to the procedure reported in the literature [101]. Compound 1d (1.85 g, 5.9 mmol) with a slight excess of LiTFSI (3.55 g, 12.4 mmol) was dissolved in 8 ml of deionized water. The solution was stirred overnight at room temperature, the solution was extracted with 30 ml of dichloromethane. The obtained organic layer was washed three times with saturated brine of 10 ml, dried over anhydrous magnesium sulfate and active carbon. The residue was purified by column chromatography on alumina and concentrated under reduced pressure to give the final product [DiMPy1O][TFSI] (yield: 3.32 g, 70.2%) as a brown liquid, and crystallized at room temperature.

IR: 3563 cm^{-1} (w), 2973 cm^{-1}, 2901 (C - H, s), 2386.94 cm^{-1}, 1931.99 - 1797.99 cm^{-1}, 1633.0 cm^{-1} 1476.73 cm^{-1}, (s = o, s), 1351.76 cm^{-1} (C - N, w), 1186.75 cm^{-1} (C - O, s), 1055.42 cm^{-1} (C - F, s), 999.41 - 876.99 cm^{-1}, 789.95 - 740.46 cm^{-1}, 645.75 - 571.02 cm^{-1}.

^{1}H NMR (300 MHz, CDCl$_3$): δ (ppm) 3.96 (t, J = 3.51 Hz, 4H), 3.85 - 3.51 (m, 12H), 3.04 (s, 6H), 2.19 (m, 8H). ^{13}C NMR (300 MHz, CDCl$_3$): δ (ppm) 119.8 (q, J = 159 Hz), 69.33, 64.41, 63.98, 61.81, 47.93, 20.84.

4.2.2.2 Synthesis of DILs [DiMPy2O][TFSI]. 1, 1' - ((ethane -1, 2 - diylbis (oxy)) bis (ethane - 2, 1 - diyl)) bis (1 - methylpyrrolidin - 1 - ium) bis ((trifluoromethyl) sulfonyl) amide: [DiMPy2O] [TFSI]. An analogous procedure was used to prepare 2d started from 2b (2.99 g, 16 mmol), 1 - methylpyrrolidine (2.86 g, 33.6 mmol), to give the desired compound as a brown liquid (2d, yield: 5.43 g, 95.1%).

The corresponding ionic liquid was prepared analogously from 2d (4.08 g, 11.4 mmol) and LiTFSI (6.49 g, 22.6 mmol) to give the final product as a brown liquid [DiMPy2O][TFSI] (yield: 6.49 g, 67.1%).

IR: 3563 cm^{-1} (w), 2964 cm^{-1}, 2902.84 (C-H, s), 2379.03 cm^{-1}, 1931.76 – 1798.58 cm^{-1} (w), 1634.43 cm^{-1}, 1463.43 cm^{-1}, (s=o, s), 1352.48 cm^{-1} (C-N, w), 1193.83 cm^{-1} (C-O, s), 1054.56 cm^{-1} (C-F, s), 953.83 – 877.41 cm^{-1}, 789.69 – 739.92 cm^{-1}.

^{1}H NMR (300 MHz, CDCl$_3$): δ (ppm) 3.84 (t, J=3.51 Hz, 4H), 3.73 – 3.50 (m, 16H), 3.03 (s, 6H), 2.17 (m, 8H). ^{13}C NMR (300 MHz, CDCl$_3$): δ (ppm) 119.8 (q, J=159 Hz), 70.2, 65.4, 65.0, 63.5, 48.6, 21.1.

4.2.2.3 Synthesis of DILs [DiMPy3O][TFSI]. 1, 1′ - (((oxybis (ethane -2, 1 - diyl)) bis (oxy)) bis (ethane -2, 1 – diyl)) bis (1 - methylpyrrolidin - 1 - ium) bis ((trifluoromethyl) sulfonyl) amide: [DiMPy3O][TFSI]. An analogous procedure was used to prepare 3d started from 3b (4.01 g, 17.3 mmol), 1 - methylpyrrolidine (3.14 g, 36.9 mmol), to give the desired compound as a dark brown liquid (3d, yield: 6.58 g, 94.5%).

The corresponding ionic liquid was prepared analogously from 3d (2.01 g, 5.0 mmol) and LiTFSI (3.02 g, 10.5 mmol) to give the final product as a dark brown liquid [DiMPy3O][TFSI], (yield: 3.17 g, 71%).

IR: 3563 cm^{-1} (w), 2973 cm^{-1}, 2877.73 (C-H, s), 2387.01 cm^{-1}, 1931.61 – 1797.99 cm^{-1} (w), 1673.96 cm^{-1}, 1462.84 cm^{-1}, (s=o, s), 1352.42 cm^{-1} (C-N, w), 1193.55 cm^{-1} (C-O, s), 1056.58 cm^{-1} (C-F, s), 998.26 – 877.85 cm^{-1}, 789.24 – 741.40

cm^{-1}.

^{1}H NMR (300 MHz, CDCl$_3$): δ (ppm) 3.87 (t, J = 3.51 Hz, 4H), 3.75 – 3.54 (m, 20H), 3.08 (s, 6H), 2.21 (m, 8H). ^{13}C NMR (300 MHz, CDCl$_3$): δ (ppm) 119.8 (q, J = 159 Hz), 70.63, 70.16, 65.50, 64.90, 48.69, 21.2.

4.2.2.4 Synthesis of DILs [DiMPy4O] [TFSI]. 1, 1' – (3, 6, 9, 12 – tetraoxatetradecane – 1, 14 – diyl) bis (1 – methylpyrrolidin – 1 – ium) bis ((trifluoromethyl) sulfonyl) amide: [DiMPy4O] [TFSI]. An analogous procedure was used to prepare 4d started from 4b (3.02 g, 11 mmol), 1 – methylpyrrolidine (2.00 g, 23.5 mmol), to give the desired compound as dark brown liquid (4d, yield: 4.02 g, 82.4%);

The corresponding ionic liquid was prepared analogously from 4d (2.66 g, 6.0 mmol) and LiTFSI (3.58 g, 12.5 mmol) to give the final product as a dark brown liquid [DiMPy4O] [TFSI] (yield: 3.97 g, 70.9%).

IR: 3542 cm^{-1} (w), 2880.44 (C – H, s), 2387.01 cm^{-1}, 1932.03 – 1723.60 cm^{-1} (w), 1673.96 cm^{-1}, 1462.71 cm^{-1}, (s = o, s), 1352.47 cm^{-1} (C – N, w), 1193.89 cm^{-1} (C – O, s), 1056.94 cm^{-1} (C – F, s), 998.59 – 839.90 cm^{-1}, 789.16 – 740.05 cm^{-1}.

^{1}H NMR (300 MHz, CDCl$_3$): δ (ppm) 3.87 (t, J = 3.51 Hz, 4H), 3.73 – 3.52 (m, 24H), 3.11 (s, 6H), 2.20 (m, 8H). ^{13}C NMR (300 MHz, CDCl$_3$): δ (ppm) 119.8 (q, J = 159 Hz), 70.3, 70.0, 65.50, 64.90, 48.7, 21.2.

4.2.2.5 Synthesis of DILs [DiMPy1O] [DCA]. Low viscous DCA-based RTILs have great potential as a co-solvent for Li rechargeable battery electrolyte. However, few research papers were published in this area, probably caused by the difficulty of the synthesis of DCA-based RTILs. The solubility of sodium dicyanamide (NaN(CN)$_2$) is pretty low in most of the organic solvents, which leads to the difficulty in the metathesis reaction. As shown in Scheme 4.2, the most commonly used method is to use silver salt as the reaction intermediate [94]. Tanner et al. reported a 24h metathesis reaction of solid-phase NaN(CN)$_2$ in acetone [102], which is rather time consuming.

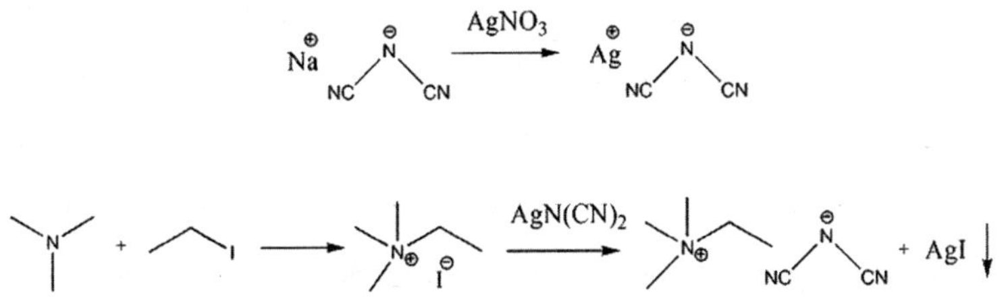

Scheme 4.2. NaN(CN)$_2$ metathesis via silver salt intermediate

We developed a concise metathesis procedure, depending on the solubility difference of NaCl and NaN(CN)$_2$ in methanol. At 25 ℃ the solubility of NaCl in methanol is 14.9 g/L [86]. We measured the solubility of NaN(CN)$_2$ at r. t. by gradually adding the salt into 30 ml of methanol, and the result is approximately 37 g/L. So for salts with chloride anion, the metathesis reaction can be carried out directly with NaN(CN)$_2$ in methanol because of the formation of NaCl precipitation, which is easily to be separated from the system.

However, similar reactions are not available for other anions including-OTs, -I or-Br. Thus, we plan to exchange the-OTs or-Br ions into-Cl by ion-exchange resin Dowex © 22 (Chloride form), which is cross-linked polymer beads, based on the styrene-divinylbenzene matrix, with plenty pending groups of dimethyl ethanol benzyl ammonium chloride. The procedure follows the reported work that produces ionic liquid by ion-exchange resin [103]. Passing the solution of our ionic liquid in methanol or MeCN through a column filled with the ion-exchange resin, the-OTs, -I or-Br will be exchanged with-Cl and absorbed in the resin, while-Cl will flow out with the cations. Dowex © 22 (Chloride form) could exchange with anions at an efficiency of 0.6 mmol/ml.

NaN(CN)$_2$ (3.71 g, 41.7 mmol) was totally dissolved in 100 ml of methanol, to which 1d (6.21 g, 19.8 mmol) with another 20 ml of methanol was added together. The mixture was stirred overnight at room temperature to remove methanol by vacuum distillation until some precipitation came out. Then the precipitation was filtered away, 15 ml of acetone was added to the filtrate. The solvent was removed by vacuum distillation until some precipitation came out again. Repeat the foregoing procedures for several times until no precipitation came out. Then the filtrate was concentrated under reduced pressure to give the final product [DiMPy1O][DCA] (yield: 5.22 g, 70.3%) as dark brown liquid and crystallized at room temperature.

IR: 3020.64 – 2891.08 (C – H, s), 2367.93 cm^{-1}, 2235.04 – 2133.44 cm^{-1} (CN, m), 1659.68 – 1461.34 cm^{-1}, 1365.82 cm^{-1} (C – N, w), 1135.15 cm^{-1} (C – O, s), 1076.29 – 987.63 cm^{-1}, 935.08 – 878.26 cm^{-1}, 817.78 – 665.65 cm^{-1}.

^{1}H NMR (300 MHz, D$_2$O): δ (ppm) 4.11 (t, J = 3.51 Hz, 4H), 3.98 – 3.71 (m, 12H), 3.23 (s, 6H), 2.34 (m, 8H). ^{13}C NMR (300 MHz, D$_2$O): δ (ppm) 119.9, 70.9, 65.3, 63.5, 48.7, 43.8, 21.4.

4.2.2.6 Synthesis of DILs [DiMPy2O][DCA]. The corresponding ionic liquid was prepared analogously from 2d (1.35, 3.8 mmol) and NaN(CN)$_2$ (1.00 g, 11.2mmol) to give the desired compound as a dark brown liquid [DiMPy2O][DCA] (yield: 1.14 g, 72%).

IR: 3023.71 – 2898.27 (C – H, s), 2365 cm^{-1}, 2238.72 – 2137.4 cm^{-1} (CN, m), 1644.48 – 1471.8 cm^{-1}, 1311.80 cm^{-1} (C – N, w),

1122.03 cm^{-1} (C-O, s), 1049.5-998.57 cm^{-1}, 935.26-905.96 cm^{-1}, 814.49 cm^{-1}.

^{1}H NMR (300 MHz, D$_2$O): δ (ppm) 3.95 (t, J = 3.51 Hz, 4H), 3.72-3.55 (m, 16H), 3.08 (s, 6H), 2.20 (m, 8H). ^{13}C NMR (300 MHz, D$_2$O): δ (ppm) 119.9, 70.8, 69.6, 64.88, 62.9, 48.6, 43.5, 21.1.

4.2.2.7 Synthesis of DILs [DiMPy3O][DCA]. The corresponding ionic liquid was prepared analogously from 3d (4.87, 12.1 mmol) and NaN(CN)$_2$ (2.28 g, 25.6 mmol) to give the desired compound as a dark brown liquid [DiMPy3O][DCA] (yield: 4.24 g, 75.7%).

IR: 3020.64-2874.89 (C-H, s), 2367 cm^{-1}, 2235.49-2135.21 cm^{-1} (CN, m), 1708.16-1462.47 cm^{-1}, 1352.87 cm^{-1} (C-N, w), 1116.4 cm^{-1} (C-O, s), 998.02-905.64 cm^{-1}, 833.73-663.06 cm^{-1}.

^{1}H NMR (300 MHz, D$_2$O): δ (ppm) 3.90 (t, J = 3.51 Hz, 4H), 3.75-3.51 (m, 20H), 3.04 (s, 6H), 2.15 (m, 8H). ^{13}C NMR (300 MHz, D$_2$O): δ (ppm) 119.9, 70.9, 69.6, 64.7, 62.9, 48.5, 43.3, 30.3, 21.1.

4.2.2.8 Synthesis of DILs [DiMpy4O][DCA]. The corresponding ionic liquid was prepared analogously from 4d (2.32 g, 5.2 mmol) and NaN(CN)$_2$ (0.98 g, 11.0 mmol) to give the desired compound as a dark brown liquid [DiMPy4O][DCA] (yield: 1.98 g, 74.9%).

IR: 3024.97-2902.28 (C-H, s), 2370.32 cm^{-1}, 2239.22-2138.81 cm^{-1} (CN, m), 1707.50-1462.18 cm^{-1}, 1312.43 cm^{-1} (C-N, w), 1111.11 cm^{-1} (C-O, s), 998.41-906.78 cm^{-1}, 833.92-665.65 cm^{-1}.

^{1}H NMR (300 MHz, D$_2$O): δ (ppm) 3.93 (t, J = 3.51 Hz, 4H),

3.72 – 3.55 (m, 24H), 3.07 (s, 6H), 2.180 (m, 8H). ^{13}C NMR (300 MHz, D_2O): δ (ppm) 119.9, 71.8, 69.6, 64.8, 63.0, 48.5, 43.4, 30.4, 21.2.

4.2.3 General procedures for the preparation of BIm-based DILs

As shown in Scheme 4.3, the synthesis of BIm-based DILs involves three steps.

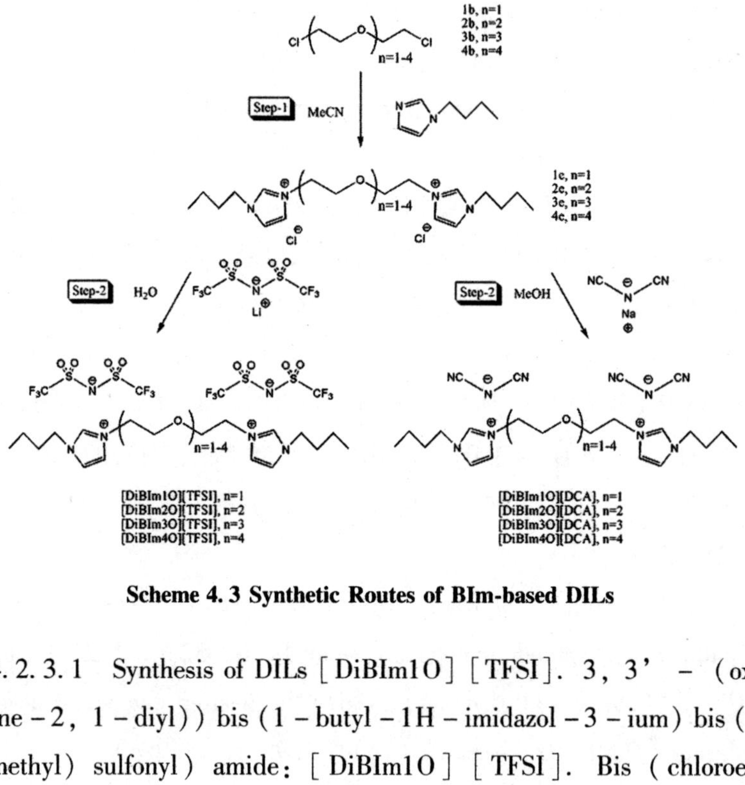

Scheme 4.3 Synthetic Routes of BIm-based DILs

4.2.3.1 Synthesis of DILs [DiBIm1O] [TFSI]. 3, 3' – (oxy bis (ethane – 2, 1 – diyl)) bis (1 – butyl – 1H – imidazol – 3 – ium) bis ((trifluoromethyl) sulfonyl) amide: [DiBIm1O] [TFSI]. Bis (chloroethyl) ether (1b, 4.09 g, 28.6 mmol) and 1 – butylimidazolium (7.48 g, 60.2 mmol) were dissolved in 10 ml of dried acetonitrile and refluxed 2h under stirring. The mixture was cooled down to room temperature and stirred overnight. After the solvent had been removed by vacuum evaporation, the residue was washed three times with 15 ml of ethyl ether. The remaining volatile was removed by high vacuum to give the desired compound 3, 3′- (oxy bis (ethane-2, 1-diyl)) bis (1-butyl-1H-imidazol-3-ium) dichloride (1e, yield: 7.85 g, 87.5%) as a brown liquid.

1e (2.44 g, 7.8 mmol) with LiTFSI (4.69 g, 16.3 mmol) was dissolved in 15 ml of deionized water. The solution was stirred overnight at room temperature, the solution was extracted with 30 ml of dichloromethane. The obtained organic layer was washed three times with saturated brine of 10 ml, dried over anhydrous magnesium sulfate and active carbon. The residue was purified by column chromatography on alumina and concentrated under reduced pressure to give the final product [DiBIm1O] [TFSI] (yield: 4.97 g, 79.6%) as a brown liquid.

IR: 3563 cm^{-1} (w), 3152.68-3117.53 cm^{-1} (= C-H, s), 2967.35-2880.04 cm^{-1} (C-H, s), 2389.19 cm^{-1}, 1932.66-1712.46 cm^{-1} (w), 1650.43 cm^{-1} (C=N, m), 1566.03-1464.08 cm^{-1}, (C=C, m), 1351.77 cm^{-1} (C-N, w), 1191.84 cm^{-1} (C-O, s), 1056.74 cm^{-1} (C-F, s), 949.51-845.29 cm^{-1}, 789.84-570.39 cm^{-1}.

^{1}H NMR (300 MHz, D-acetone): δ (ppm) 9.18 (s, 2H), 7.85 (d, J=1.2 Hz, 2H), 7.72 (d, J=1.2 Hz, 2H), 4.55 (t, J=4.5 Hz, 4H), 4.34 (t, J=7.2 Hz, 4H), 3.97 (t, J=6.9 Hz, 4H), 1.90 (m, 4H), 1.37 (m, 4H), 0.92 (t, J=7.2 Hz, 6H). ^{13}C NMR (300 MHz, D-acetone): δ (ppm) 135.5, 123.2, 122.2, 119.8 (q, J=159 Hz), 68.7, 49.5, 49.4, 31.8, 19.3, 12.7.

4.2.3.2 Synthesis of DILs [DiBIm2O] [TFSI]. 3, 3′- ((ethane-1, 2-diylbis (oxy)) bis (ethane-2, 1-diyl)) bis (1-butyl-1H-imidazol-3-ium) bis ((trifluoromethyl) sulfonyl) amide: [DiBIm2O] [TFSI]. An analogous procedure was used to prepare 2e started from 2b

(2.93 g, 15.6 mmol), 1 - butylimidazolium (4.09 g, 32.9 mmol), to give the desired compound as a brown liquid (2e, yield: 5.01 g, 89.7%).

The corresponding ionic liquid was prepared analogously from 2e (1.97 g, 5.5 mmol) and LiTFSI (3.32 g, 11.5 mmol) to give the final product as a brown liquid [DiBIm2O][TFSI] (yield: 3.74 g, 80.3%).

IR: 3563 cm^{-1} (w), 3152.04 - 3116.51 cm^{-1} (= C-H, s), 2966.46 - 2879.14 cm^{-1} (C - H, s), 2389.30 cm^{-1}, 1932.15 - 1712.48 cm^{-1} (w), 1650.43 cm^{-1} (C = N, m), 1565.96 - 1463.53 cm^{-1}, (C = C, m), 1352.18 cm^{-1} (C - N, w), 1191.82 cm^{-1} (C - O, s), 1056.64 cm^{-1} (C - F, s), 929.50 - 834.69 cm^{-1}, 789.62 - 559.98 cm^{-1}.

^{1}H NMR (300 MHz, CDCl$_3$): δ (ppm) 8.7 (s, 2H), 7.46 (d, J = 1.2 Hz, 2H), 7.4 (d, J = 1.2 Hz, 2H), 4.34 (t, J = 4.5 Hz, 4H), 4.17 (t, 6.9 Hz, 4H), 3.79 (t, J = 5.1 Hz, 4H), 3.48 (t, J = 3.9 Hz, 4H), 1.83 (m, 4H), 1.2 (m, 4H), 0.91 (t, 6H). ^{13}C NMR (300 MHz, CDCl$_3$): δ (ppm) 135.5, 123.2, 122.2, 119.8 (q, J = 159 Hz), 70.2, 68.6, 50.8, 49.8, 31.8, 19.2, 13.0.

4.2.3.3 Synthesis of DILs [DiBIm3O][TFSI]. 3, 3' - (((oxybis (ethane - 2, 1 - diyl)) bis (oxy)) bis (ethane - 2, 1 - diyl)) bis (1 - butyl - 1H - imidazol - 3 - ium) bis ((trifluoromethyl) sulfonyl) amide: [DiBIm3O][TFSI]. An analogous procedure was used to prepare 3e started from 3b (3.78 g, 16.4 mmol), 1 - butylimidazolium (4.28 g, 34.4 mmol), to give the desired compound as a brown liquid (3e, yield: 5.86 g, 89.2%).

The corresponding ionic liquid was prepared analogously from 3e (2.74 g, 6.8 mmol) and LiTFSI (4.11 g, 14.3 mmol) to give the final product as a brown liquid [DiBIm3O][TFSI] (yield: 4.78g, 78.7%).

IR: 3563 cm^{-1} (w), 3151.15 - 3115.88 cm^{-1} (= C-H, s), 2965.73 - 2878.65 cm^{-1} (C - H, s), 2387.36 cm^{-1}, 1929.38 - 1714.20 cm^{-1} (w), 1650.43 cm^{-1} (C = N, m), 1565.61 - 1464.09 cm^{-1}, (C = C, m), 1351.86 cm^{-1} (C - N, w), 1191.84 cm^{-1} (C - O, s), 1057.15 cm^{-1} (C - F, s), 936.01 - 838.87 cm^{-1}, 789.4 - 570.62 cm^{-1}.

^{1}H NMR (300 MHz, D-acetone): δ (ppm) 9.04 (s, 2H), 7.74 (d, J = 1.2 Hz, 2H), 7.72 (d, J = 1.2 Hz, 2H), 4.50 (t, J = 4.2 Hz, 4H), 4.35 (t, J = 6.9 Hz, 4H), 3.91 (t, J = 4.2 Hz, 4H), 3.65 - 3.20 (t, J = 3.9 Hz, 8H), 1.90 (m, 4H), 1.37 (m, 4H), 0.92 (t, J = 7.2 Hz, 6H). ^{13}C NMR (300 MHz, D-acetone): δ (ppm) 135.5, 123.2, 122.2, 119.8 (q, J = 159 Hz), 70.1, 68.5, 49.6, 49.4, 31.8, 19.1, 12.8.

4.2.3.4 Synthesis of DILs [DiBIm4O] [TFSI]. 3, 3′ - (3, 6, 9, 12 - tetraoxatetradecane - 1, 14 - diyl) bis (1 - butyl - 1H-imidazol - 3 - ium) bis ((trifluoromethyl) sulfonyl) amide: [DiBIm4O] [TFSI]. An analogous procedure was used to prepare 4e started from 4b (2.95 g, 10.7 mmol), 1 - butylimidazolium (2.83 g, 22.8 mmol), to give the desired compound as a brown liquid (4e, yield: 4.28 g, 89.6%).

The corresponding ionic liquid was prepared analogously from 4e (1.93 g, 4.3 mmol) and LiTFSI (2.61 g, 9.1 mmol) to give the final product as a brown liquid [DiBIm4O] [TFSI] (yield: 3.23 g, 79.8%).

IR: 3542.72 cm^{-1} (w), 3149.39 - 3114.79 cm^{-1} (= C-H, s), 2963.48 - 2877.37 cm^{-1} (C - H, s), 2389.19 cm^{-1}, 1926.6 - 1846.11 cm^{-1} (w), 1650.43 cm^{-1} (C = N, m), 1565.6 - 1463.43 cm^{-1}, (C = C, m), 1352.27 cm^{-1} (C - N, w), 1191.84 cm^{-1} (C - O, s), 1057.7 cm^{-1} (C - F, s), 939.86 - 840.24 cm^{-1}, 789.03 - 570.54

cm^{-1}.

^{1}H NMR (300 MHz, D-acetone): δ (ppm) 9.05 (s, 2H), 7.75 (d, J = 1.2 Hz, 2H), 7.73 (d, J = 1.2 Hz, 2H) 4.49 (t, J = 4.5 Hz, 4H), 4.35 (t, J = 7.2 Hz, 4H), 3.88 (t, J = 4.8 Hz, 4H), 3.71 – 3.37 (t, J = 1.8 Hz, 12H), 1.90 (m, 4H), 1.37 (m, 4H), 0.92 (t, J = 7.2 Hz, 6H). ^{13}C NMR (300 MHz, D-acetone): δ (ppm) 135.5, 123.2, 122.2, 119.8 (q, J = 159 Hz), 70.2, 68.4, 49.6, 49.4, 31.8, 19.1, 12.8.

4.2.3.5 Synthesis of DILs [DiBIm1O][DCA]. $NaN(CN)_2$ (1.36 g, 15.3 mmol) was totally dissolved in 40 ml of methanol, to which 1e (, 2.28 g, 7.3 mmol) with another 5 ml methanol was added together. The mixture was stirred overnight at room temperature, and the methanol was removed by vacuum distillation until some precipitation came out. Then the precipitation was filtered away, 15 ml of acetone was added into the filtrate. The solvent was removed by vacuum distillation until some precipitation came out again. Repeat the foregoing procedures for several times until no precipitation came out. Then the filtrate was concentrated under reduced pressure to give the final product [DiBIm1O][DCA] (yield: 2.03 g, 74.7%) as a dark brown liquid.

IR: 3144.59 – 3106.59 cm^{-1} (=C-H, s), 2962.44 – 2875.54 cm^{-1} (C – H, s), 2240.53 – 2138.22 cm^{-1} (CN, m), 1633.47 cm^{-1} (C = N, m), 1565.21 – 1463.45 cm^{-1}, (C = C, m), 1316.78 cm^{-1} (C – N, w), 1165.31 – 1129.01 cm^{-1} (C – O, s), 907.02 – 752.42 cm^{-1}.

^{1}H NMR (300 MHz, D_2O): δ (ppm) 8.74 (s, 2H), 7.46 (d, J = 1.2 Hz, 2H), 7.42 (d, J = 1.2 Hz, 2H), 4.34 (t, J = 4.5 Hz, 4H), 4.15 (t, J = 7.2 Hz, 4H), 3.83 (t, J = 4.8 Hz, 4H), 1.8 (m, 4H), 1.23 (m, 4H), 0.86 (t, J = 7.2 Hz, 6H). ^{13}C NMR (300 MHz, D_2O):

δ (ppm) 135.6, 122.8, 122.3, 119.7, 68.6, 49.5, 31.4, 30.3, 18.9, 12.8.

4.2.3.6 Synthesis of DILs [DiBIm2O][DCA]. The corresponding ionic liquid was prepared analogously from 2e (2.98 g, 8.3 mmol) and NaN(CN)$_2$ (1.56 g, 17.5 mmol) to give the desired compound as a dark brown liquid [DiBIm2O][DCA] (yield: 2.61 g, 74.8%).

IR: 3143.53 – 3105.28 cm^{-1} (=C-H, s), 2962.23 – 2875.06 cm^{-1} (C – H, s), 2236.83 – 2135.58 cm^{-1} (CN, m), 1644.07 cm^{-1} (C = N, m), 1565.05 – 1464.30 cm^{-1}, (C = C, m), 1313.82 cm^{-1} (C – N, w), 1165.47 – 1116.54 cm^{-1} (C – O, s), 906.6 – 752.89 cm^{-1}.

^{1}H NMR (300 MHz, D$_2$O): δ (ppm) 8.74 (s, 2H), 7.48 (d, J = 1.2 Hz, 2H), 7.47 (d, J = 1.2 Hz, 2H), 4.33 (t, J = 4.5 Hz, 4H), 4.15 (t, J = 6.9 Hz, 4H), 3.82 (t, J = 5.1 Hz, 4H), 3.60 (t, 4H), 1.8 (m, 4H), 1.24 (m, 4H), 0.85 (t, J = 7.2 Hz, 6H). ^{13}C NMR (300 MHz, D$_2$O): δ (ppm) 135.7, 122.7, 122.3, 119.7, 69.7, 68.5, 49.5, 31.4, 30.3, 18.9, 12.8.

4.2.3.7 Synthesis of DILs [DiBIm3O][DCA]. The corresponding ionic liquid was prepared analogously from 3e (2.12 g, 5.3 mmol) and NaN(CN)$_2$ (0.99 g, 11.1 mmol) to give the desired compound as a dark brown liquid [DiBIm3O][DCA] (yield: 1.85 g, 75.8%).

IR: 3143.50 – 3105.37 cm^{-1} (=C-H, s), 2961.67 – 2874.75 cm^{-1} (C – H, s), 2237.71 – 2136.15 cm^{-1} (CN, m), 1644.40 cm^{-1} (C = N, m), 1564.91 – 1463.74 cm^{-1}, (C = C, m), 1314.05 cm^{-1} (C – N, w), 1165.26 – 1116.69 cm^{-1} (C – O, s), 907.06 – 752.89 cm^{-1}.

^{1}H NMR (300 MHz, D$_2$O): δ (ppm) 8.74 (s, 2H), 7.51 (d, J = 1.2 Hz, 2H), 7.48 (d, J = 1.2 Hz, 2H), 4.36 (t, J = 4.2 Hz, 4H), 4.17 (t, J = 6.9 Hz, 4H), 3.85 (t, J = 4.2 Hz, 4H), 3.59 – 3.64 (t,

J = 3.9 Hz, 8H), 1.8 (m, 4H), 1.23 (m, 4H), 0.86 (t, J = 7.2 Hz, 6H). ^{13}C NMR (300 MHz, D_2O): δ (ppm) 135.7, 122.8, 122.3, 119.7, 69.8, 69.6, 68.5, 49.5, 31.4, 30.3, 19.0, 12.9.

4.2.3.8 Synthesis of DILs [DiBIm4O] [DCA]. The corresponding ionic liquid was prepared analogously from 4e (2.35 g, 5.3 mmol) and NaN(CN)$_2$ (0.99 g, 11.1 mmol) to give the desired compound as a dark brown liquid [DiBIm4O] [DCA] (yield: 2.04 g, 76.2%).

IR: 3143.97 – 3107.37 cm^{-1} (=C-H, s), 2961.01 – 2875.06 cm^{-1} (C – H, s), 2239.71 – 2138.17 cm^{-1} (CN, m), 1633.23 cm^{-1} (C = N, m), 1565.11 – 1464.06 cm^{-1}, (C = C, m), 1315.39 cm^{-1} (C – N, w), 1114.62 cm^{-1} (C – O, s), 945.74 – 752.33 cm^{-1}.

^{1}H NMR (300 MHz, D_2O): δ (ppm) 8.74 (s, 2H), 7.48 (d, J = 1.2 Hz, 2H), 7.42 (d, J = 1.2 Hz, 2H), 4.3 (t, J = 4.5 Hz, 4H), 4.11 (t, J = 7.2 Hz, 4H), 3.8 (t, J = 4.8 Hz, 4H), 3.56 – 3.61 (t, J = 1.8 Hz, 12H), 1.8 (m, 4H), 1.23 (m, 4H), 0.86 (t, J = 7.2 Hz, 6H). ^{13}C NMR (300 MHz, D_2O): δ (ppm) 135.7, 122.7, 122.2, 119.8, 71.7, 70.8, 69.7, 68.5, 49.4, 31.8, 30.3, 19.0, 12.7.

4.2.4 General procedures for the preparation of DES-based DILs

As shown in Scheme 4.4, the synthesis of DES-based DILs along with TFSI anions involves three steps, and DCA anions reactions involve three steps.

4.2.4.1 Synthesis of DILs [DiDES1O] [TFSI]. (oxy bis (ethane – 2, 1 – diyl)) bis (diethyl sulfonium) bis ((trifluoromethyl) sulfonyl) amide: [DiDES1O] [TFSI]. Diethylene glycol 1a (4.01 g, 37.8 mmol) and 4 – methylbenzenesulfonyl chloride (TsCl) (15.08 g, 79.4 mmol) were dissolved in 50 ml of DCM. KOH (6.05 g, 108 mmol) was added slowly into the solution with ice-water bath under vigorous stirring. After stirring over-

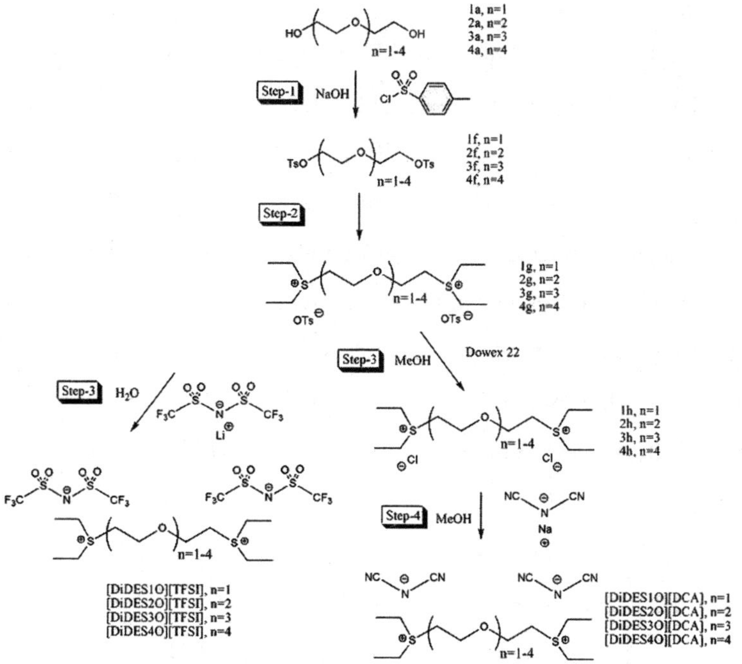

Scheme 4.4. Synthetic Routes of DES-based DILs.

night at room temperature, 50 ml of DCM was added to dilute the mixture, and then it was washed with 50 ml of water one time and 20 ml of brine twice. The organic layer was separated and dried with anhydrous $MgSO_4$. Then the solvent was removed to get the desired product. (oxy bis (ethane - 2, 1 - diyl)) bis (4 - methylbenzenesulfonate): (1f, Yield: 14.56 g, 93%), which was a colorless viscous liquid and crystallized at room temperature.

1f (10.15 g, 24.5 mmol) and diethyl sulfide (8.88 g, 98.4mmol) were mixed in a neat flask and refluxed for two days. The remains were washed with 15 mL of ethyl ether three times and dried under high vacuum. The pale yellow viscous liquid product (oxy bis (ethane - 2, 1 - diyl)) bis (diethyl sulfonium) 4 - methylbenzenesulfonate: (1g, yield: 11.16 g,

76.6%) was attained.

1g (4.54 g, 7.6 mmol) and LiTFSI (4.8 g, 16.7 mmol) were dissolved in 10 ml of deionized water. The solution was refluxed overnight at 80 ℃, the solution was extracted with 30 ml of dichloromethane. The obtained organic layer was washed three times with saturated brine of 10 ml, dried over anhydrous magnesium sulfate. The residue was purified by column chromatography on alumina and concentrated under reduced pressure to give the final product [DiDES1O][TFSI] (Yield: 4.43g, 71.5%) as a pale liquid.

IR: $2987.41 - 2885.08$ cm^{-1} (C-H, s), $1934.93 - 1790.59$ cm^{-1} (w), $1456.03 - 1351.84$ cm^{-1} (C-H, m), 1189.75 cm^{-1} (C-O, s), $1056.67 - 977.13$ cm^{-1} (C-F, s), $789.49 - 740.25$ cm^{-1}, $654 - 570.4$ cm^{-1}.

^{1}H NMR (300 MHz, CDCl$_3$): δ (ppm) 3.95 (t, J = 4.2 Hz, 4H), 3.45 (t, J = 2.4 Hz, 4H), 3.23 (m, 8H), 1.46 (t, J = 7.2 Hz, 12H). ^{13}C NMR (300 MHz, CDCl$_3$): δ (ppm) 119.8 (q, J = 159 Hz), 65.2, 33.8, 32.8, 8.5.

4.2.4.2 Synthesis of DILs [DiDES2O][TFSI]. ((ethane-1, 2-diylbis(oxy)) bis (ethane-2, 1-diyl)) bis (diethylsulfonium) bis ((trifluoromethyl) sulfonyl) amide: [DiDES2O][TFSI]. An analogous procedure was used to prepare 2f (8.03g, 86.8%) started with 2a (3.03 g, 20.2 mmol), KOH (4.52 g, 80.6 mmol) and TsCl (8.42 g, 44.3 mmol). Then 2f (6.03, 13.1 mmol) was taken to react with DES (4.75 g, 52.7) to give the desired compound as a yellow liquid (2g, yield: 6.85 g, 81.6%).

The corresponding ionic liquid was prepared analogously from 2g (3.57 g, 5.6 mmol) and LiTFSI (3.56 g, 12.4 mmol) to give the final product as a yellow liquid [DiDES2O][TFSI] (yield: 3.64 g, 76%).

IR: 2987.54 – 2881.83 cm^{-1} (C–H, s), 1456.03 – 1351.84 cm^{-1} (C–H, m), 1189.75 cm^{-1} (C–O, s), 1056.67 – 977.13 cm^{-1} (C–F, s), 789.49 – 740.25 cm^{-1}, 654 – 570.39 cm^{-1}.

^{1}H NMR (300 MHz, CDCl$_3$): δ (ppm) 3.93 (t, J = 5.1 Hz, 4H), 3.59 (t, J = 4.2 Hz, 4H), 3.46 (t, J = 2.4 Hz, 4H), 3.29 (m, 8H), 1.44 (t, J = 7.2 Hz, 12H). ^{13}C NMR (300 MHz, CDCl$_3$): δ (ppm) 119.8 (q, J = 159 Hz), 70.3, 65.1, 34.2, 32.7, 8.5.

4.2.4.3 Synthesis of DILs [DiDES3O][TFSI]. (((oxybis(ethane-2,1-diyl))bis(oxy))bis(ethane-2,1-diyl))bis(diethylsulfonium) bis((trifluoromethyl)sulfonyl)amide: [DiDES3O][TFSI]. An analogous procedure was used to prepare 3f (10.59 g, 89.2%) started from 3a (4.59 g, 23.6 mmol), KOH (5.3 g, 94.5 mmol) and TsCl (9.87 g, 52 mmol). Then 3f (9.59 g, 19.1 mmol) was taken to react with DES (6.88 g, 76.3 mmol) to give the desired compound as a yellow liquid (3g, yield: 9.87 g, 75.7%).

The corresponding ionic liquid was prepared analogously from 3g (3.05 g, 4.5 mmol) and LiTFSI (2.78 g, 9.7 mmol) to give the final product as a yellow liquid [DiDES3O][TFSI] (yield: 3.11 g, 69.8%).

IR: 2992.44 – 2879.84 cm^{-1} (C–H, s), 1934.93 – 1793.37 cm^{-1} (w), 1455.73 – 1351.99 cm^{-1} (C–H, m), 1190.57 cm^{-1} (C–O, s), 1057.11 – 976.06 cm^{-1} (C–F, s), 789.49 – 740.25 cm^{-1}, 663.24 – 555.13 cm^{-1}.

^{1}H NMR (300 MHz, CDCl$_3$): δ (ppm) 3.94 (t, J = 5.1 Hz, 4H), 3.7 – 3.5 (t, J = 4.2 Hz, 8H), 3.46 (t, J = 2.4 Hz, 4H), 3.23 (m, 8H), 1.46 (t, J = 7.2 Hz, 12H). ^{13}C NMR (300 MHz, CDCl$_3$): δ (ppm) 119.8 (q, J = 159 Hz), 70.5, 70.1, 65.3, 34.2, 32.8, 8.6.

4.2.4.4 Synthesis of DILs [DiDES4O][TFSI]. (3,6,9,12-tetraoxatetradecane – 1, 14 – diyl) bis (diethylsulfonium) bis ((trifluoromethyl) sulfonyl) amide：[DiDES4O][TFSI]. An analogous procedure was used to prepare 4f (10.11 g, 82.8%) started from 4a (4.87 g, 20.4 mmol), KOH (4.59 g, 81.8 mmol) and TsCl (8.54 g, 45 mmol). Then 4f (9.18 g, 16.8 mmol) was taken to react with DES (6.1 g, 67.6 mmol) to give the desired compound as a dark yellow liquid (4g, yield：10.11 g, 82.8%).

The corresponding ionic liquid was prepared analogously from 4g (2.58 g, 3.6 mmol) and LiTFSI (2.22 g, 7.7 mmol) to give the final product as a dark yellow liquid [DiDES4O][TFSI] (yield：2.54 g, 75.7%).

IR：2946.74 – 2875.87 cm^{-1} (C – H, s), 1931.78 – 1798.10 cm^{-1} (w), 1456.54 – 1351.42 cm^{-1} (C – H, m), 1189.75 cm^{-1} (C – O, s), 1056.67 – 977.13 cm^{-1} (C – F, s), 789.49 – 740.25 cm^{-1}, 654 – 570.39 cm^{-1}.

^{1}H NMR (300 MHz, $CDCl_3$)：δ (ppm) 3.92 (t, J = 2.1 Hz, 4H), 3.7 – 3.51 (t, J = 5.1 Hz, 12H), 3.43 (t, J = 5.7 Hz, 4H), 3.18 (m, 8H), 1.36 (t, J = 6.6 Hz, 12H). ^{13}C NMR (300 MHz, $CDCl_3$)：δ (ppm) 119.8 (q, J = 159 Hz), 71.1, 70.4, 69.5, 65.1, 34.2, 32.8, 8.5.

4.2.4.5 Synthesis of DILs [DiDES1O][DCA]. (oxy bis (ethane – 2, 1 – diyl)) bis (diethyl sulfonium) dicyanamide：[DiDES1O][DCA]. Dowex ⓒ 22 (Chloride form) resin was washed three times with methanol and dried in the oven under 60 ℃ to remove all moisture. Dried Dowex ⓒ 22 of 20 ml contained in a column and wetted with 15 mL of methanol, 1g (6.42 g, 10.8 mmol) was dissolved in 15 mol of methanol. This solution was passed through the resin column with a dropping rate controlled to ~1 drop/second. Methanol of 60mL was added to wash the column. Then all methanol solutions were combined together and the solvent was removed. The desired product (oxy bis (ethane -2, 1 - diyl)) bis (diethyl sulfonium) dichloride 1h (Yield: 2.97g, 85.1%) was obtained as a brown viscous liquid.

NaN(CN)$_2$ (1.39g, 15.6 mmol) was totally dissolved in 40ml of methanol, 1h (2.37 g, 7.3 mmol) with another 5 ml methanol was added together. The mixture was stirred overnight at room temperature, the methanol was removed by vacuum distillation until some precipitation came out. Then the precipitation was filtered away, 15 ml of acetone was added to the filtrate. The solvent was removed by vacuum distillation until some precipitation came out again. repeat the preceding procedures for several times until no precipitation came out. Then the filtrate was concentrated under reduced pressure to give the final product [DiDES1O] [DCA] (yield: 2.1 g, 74.3%) as a yellow liquid.

IR: 2987.41 - 2877.08 cm^{-1} (C-H, s), 2236.96 - 2135.93 cm^{-1} (CN, m), 1650.71 cm^{-1} (C=N, m), 1565.11 - 1422.63 cm^{-1}, (C=C, m), 1310.94 cm^{-1} (C-N, w), 1112.76 cm^{-1} (C-O, s), 976.20 - 790.97 cm^{-1}, 665.31 cm^{-1}.

^{1}H NMR (300 MHz, CDCl$_3$): δ (ppm) 3.97 (t, J=4.2 Hz, 4H), 3.64 (t, J=2.4 Hz, 4H), 3.4 - 3.18 (m, 8H), 1.46 (t, J=7.2 Hz, 12H). ^{13}C NMR (300 MHz, CDCl$_3$): δ (ppm) 119.9, 65.2, 33.8, 32.8, 8.5.

4.2.4.6 Synthesis of DILs [DiDES2O] [DCA]. ((ethane -1, 2 - diylbis (oxy)) bis (ethane -2, 1 - diyl)) bis (diethylsulfonium) dicyanamide: [DiDES2O] [DCA]. An analogous procedure was used to prepare 2h started with 2g (3.09 g, 4.8 mmol) to give the desired compound as a yellow liquid (2h, yield: 1.55 g, 87.2%);

The corresponding ionic liquid was prepared analogously from 2h (1.15

g, 3.1 mmol) and NaN(CN)$_2$ (0.36 g, 4.1 mmol) to give the final product as a yellow liquid [DiDES2O][DCA] (yield: 1.01 g, 75.7%).

IR: 2979.21 – 2877.82 cm^{-1} (C – H, s), 2238.92 – 2137.95 cm^{-1} (CN, m), 1650.89 cm^{-1} (C = N, m), 1581.58 – 1455.26 cm^{-1}, (C = C, m), 1313.77 cm^{-1} (C – N, w), 1119.62 cm^{-1} (C – O, s), 1034.44 – 790.97 cm^{-1}, 665.31 cm^{-1}.

^{1}H NMR (300 MHz, CDCl$_3$): δ (ppm) 3.93 (t, J = 5.1 Hz, 4H), 3.59 (t, J = 4.2 Hz, 4H), 3.46 (t, J = 2.4 Hz, 4H), 3.29 – 3.19 (m, 8H), 1.36 (t, J = 7.2 Hz, 12H). ^{13}C NMR (300 MHz, CDCl$_3$): δ (ppm) 119.9, 69.9, 64.7, 33.5, 32.0, 8.0.

4.2.4.7　Synthesis of DILs [DiDES3O][DCA]. (((oxybis(ethane-2,1-diyl))bis(oxy))bis(ethane-2,1-diyl))bis(diethylsulfonium) dicyanamide: [DiDES3O][DCA]. An analogous procedure was used to prepare 3h started with 3g (6.72 g, 9.8 mmol) to give the desired compound as a yellow liquid (3h, yield: 3.38 g, 83.4%).

The corresponding ionic liquid was prepared analogously from 3h (2.48 g, 6.0 mmol) and NaN(CN)$_2$ (1.13 g, 12.6 mmol) to give the final product as a yellow liquid [DiDES3O][DCA] (yield: 2.16 g, 75.8%).

IR: 3029.41 – 2877.08 cm^{-1} (C – H, s), 2248.57 – 2139.07 cm^{-1} (CN, m), 1650.61 cm^{-1} (C = N, m), 1581.72 – 1397.34 cm^{-1}, (C = C, m), 1318.19 cm^{-1} (C – N, w), 1121.96 cm^{-1} (C – O, s), 1033.73 – 790.11 cm^{-1}, 665.37 cm^{-1}.

^{1}H NMR (300 MHz, CDCl$_3$): δ (ppm) 3.93 (t, J = 5.1 Hz, 4H), 3.8 – 3.53 (t, J = 4.2 Hz, 8H), 3.45 (t, J = 2.4 Hz, 4H), 3.3 – 3.23 (m, 8H), 1.39 (t, J = 7.2 Hz, 12H). ^{13}C NMR (300 MHz, CDCl$_3$): δ (ppm) 119.9, 70.0, 69.6, 64.7, 38.6, 33.7, 8.1.

4.2.4.8　Synthesis of DILs [DiDES4O][DCA]. (3,6,9,12 – tetraoxatetradecane − 1, 14 − diyl) bis (diethylsulfonium) dicyanamide: [DiDES4O][DCA]. An analogous procedure was started with 4g (7.43 g, 10.2 mmol) to give the desired compound as a yellow liquid (4h, yield: 3.81 g, 81.8%).

The corresponding ionic liquid was prepared analogously from 4h (2.81 g, 6.2 mmol) and NaN(CN)$_2$ (1.15 g, 13 mmol) to give the final product as a yellow liquid liquid [DiDES4O][DCA] (yield: 2.32 g, 72.6%).

IR: 2987.41 − 2877.08 cm^{-1} (C − H, s), 2236.96 − 2135.93 cm^{-1} (CN, m), 1650.71 cm^{-1} (C = N, m), 1565.11 − 1422.63 cm^{-1}, (C = C, m), 1310.94 cm^{-1} (C − N, w), 1112.76 cm^{-1} (C − O, s), 976.20 − 790.97 cm^{-1}, 665.31 cm^{-1}.

^{1}H NMR (300 MHz, D$_2$O): δ (ppm) 3.95 (t, J = 2.1 Hz, 4H), 3.8 − 3.53 (t, J = 5.1 Hz, 12H), 3.45 (t, J = 5.7 Hz, 4H), 3.3 − 3.18 (m, 8H), 1.6 (t, J = 6.6 Hz, 12H). ^{13}C NMR (300 MHz, D$_2$O): δ (ppm) 120, 69.9, 69.7, 69.4, 64.7, 33.6, 32.1, 8.1.

4.3 Results and Discussion

4.3.1 Characterization

In order to make the best choice of the electrolyte composition for use in a practical Li rechargeable battery, we should take into account the flammability, conductivity, and viscosity of the electrolytes [104]. Diethylene glycol, triethylene glycol, tetraethylene glycol, pentaethylene glycol with different lengths of EO chains were chosen as the ideal compounds, because of their structural similarity and tunability (i. e., the ability to alter, control, and tailor their structure as desired), contributed to understanding better the structural change that corresponds to a particular physical property. They were modified by converting the end groups to an appropriate leaving group such as chloride. The subsequent quaternization reaction with MPy (or BIm) at elevated temperature (reflux) led to the corresponding DILs with Cl$^-$ anions, whereas TFSI (or DCA) was obtained via further anion exchange. However, applying the same procedure for the preparation of DES-based DILs did not result in the products of a good yield. In order to obtain the product smoothly, efficiently, we changed the strategy by replacing of the dichloride precursor with the more active di-OTs compound and extending the reaction time up to 72 h. Finally, we got the desired products. All the physicochemical properties of the DILs are listed in Table 4.1.

To make the best choice of the electrolyte composition for use in a practical Li rechargeable battery, we should take into account the flammability, conductivity, and viscosity of the electrolytes [104]. Diethylene glycol, triethylene glycol, tetraethylene glycol, pentaethylene glycol with different length of EO chains were chosen as the ideal compounds, because of their structural similarity and tunability (i. e., the ability to alter, control, and tailor their structure as desired), contributed to understanding better the structural change that corresponds to a particular physical property. They were modified by converting the end groups to an appropriate leaving group such as chloride. The subsequent quaternization reaction with MPy (or BIm) at elevated temperature (reflux) led to the corresponding DILs with Cl-anions, whereas TFSI (or DCA) was obtained via further anion exchange. However, applying the same procedure for the preparation of DES-based DILs did not result in the products of a good yield. In order to obtain the product smoothly, efficiently, we changed the strategy by replacing of the dichloride precursor with the more active di-OTs compound and extending the reaction time up to 72 h, we got the desired products finally.

Table 4.1. Physicochemical properties of the DILs

Compounds		Molecular weight (g/mol)	Density (g/ml)[a]	Viscosity (cp)[b]	Appearance	Ionic conductivity Neat (mS/cm)	
						25 ℃	70 ℃
[MImEt]	[DCA]	177.21	1.10	14.82	pale yellow liquid	26.3	62.3
[MImEt]	[TFSI]	391.31	1.47	25	colorless liquid	10.62	27.1
[DiMPy1O]	[DCA]	374.48	1.19	>2000	dark brown crystal	0.439	8.12
[DiMPy2O]	[DCA]	418.54	1.21	64.31	brown liquid	2.3	12.95
[DiMPy3O]	[DCA]	462.59	1.22	60.44	dark brown liquid	2.5	12.93
[DiMPy4O]	[DCA]	506.64	1.22	49.49	dark brown liquid	1.445	9.47
[DiMPy1O]	[TFSI]	802.69	1.56	>2000	brown crystal	1.23	7.39
[DiMPy2O]	[TFSI]	846.75	1.57	377.5	brown liquid	1.4	6.01
[DiMPy3O]	[TFSI]	890.8	1.59	279	dark brown liquid	1.01	4.9
[DiMPy4O]	[TFSI]	934.85	1.61	189.6	dark brown liquid	0.971	4.54
[DiBIm1O]	[DCA]	452.26	1.21	429.2	dark brown liquid	2.31	14.7
[DiBIm2O]	[DCA]	496.61	1.21	169	dark brown liquid	1.532	10.16
[DiBIm3O]	[DCA]	540.66	1.21	134.9	dark brown liquid	1.696	10.11
[DiBIm4O]	[DCA]	584.71	1.22	75.33	dark brown liquid	1.356	11.31
[DiBIm1O]	[TFSI]	880.77	1.51	411.4	brown liquid	0.646	4.56
[DiBIm2O]	[TFSI]	924.82	1.53	364	brown liquid	0.785	5.26
[DiBIm3O]	[TFSI]	968.87	1.54	332.8	brown liquid	0.774	4.74
[DiBIm4O]	[TFSI]	1012.92	1.57	298.6	brown liquid	0.721	4.75
[DiDES1O]	[DCA]	384.56	1.25	368.5	yellow liquid	1.868	13.1
[DiDES2O]	[DCA]	428.62	1.24	361.1	yellow liquid	0.674	5.39
[DiDES3O]	[DCA]	472.67	1.24	142.9	yellow liquid	1.43	11.74
[DiDES4O]	[DCA]	516.72	1.23	85.22	yellow liquid	3.41	16.8
[DiDES1O]	[TFSI]	812.77	1.60	366.1	pale yellow liquid	2.47	10.28
[DiDES2O]	[TFSI]	856.83	1.57	235.7	yellow liquid	0.885	4.48
[DiDES3O]	[TFSI]	900.88	1.54	128.1	yellow liquid	1.27	7.54

continued

Compounds	Molecular weight (g/mol)	Density (g/ml)[a]	Viscosity (cp)[b]	Appearance	Ionic conductivity Neat (mS/cm)	
					25 ℃	70 ℃
[DiDES4O][TFSI]	944.93	1.53	110.4	dark yellow liquid	2.4	11.55

a, b was measured at 25 ℃

Applying the presented synthesis route a series of DILs with various cationic head groups and different anionic counterparts could be obtained with high yields. Furthermore, using this method, we have the ability to control and tune the properties of DILs to get the desirable solvation properties and thermal stabilities [105].

Three series of DILs with spacer lengths from EO1 to EO4 have been synthesized, and some of their properties such as density, viscosity, and appearance at room temperature have been investigated [36]. All the physicochemical properties of the DILs are listed in Table 4.1. Compounds [DiMPy1O][DCA] and [DiMPy1O][TFSI] with one EO chain spacer length, respectively, were crystallized at room temperature, whereas other compounds with spacer length (EO chains $\geq$ 2) were liquids.

The density is the most often measured and reported physical property of ionic liquids because nearly every application requires knowledge of the density. Table 4.1 contains density values of ILs, in general, ILs are denser than either organic solvents and water. The density properties follow quite consistent trends. For a given cation, the density increases as the molecular weight of the anion increases, just as DILs containing TFSI anions have a higher density than those containing DEA anions. For instance, [DiMPy1O][TFSI] (1.5568 g cm^{-1} at 25℃) has higher density than [DiMPy1O][DCA] (1.19 g cm^{-1} at 25 ℃), [DiBIm1O][TFSI] (1.51 g cm^{-1} at 25℃) has higher density than [DiBIm1O][DCA] (1.21 g cm^{-1} at 25℃),

[DiDES1O] [TFSI] (1.60 g cm^{-1} at 25℃) has higher density than [DiDES1O] [DCA] (1.25 g cm^{-1} at 25℃). Since none of the anions have a particularly long or awkward chain, it is possible that they all can occupy favorable, relatively close approach positions around the cation, resulting in the high densities [106].

On the other hand, MPy-based DILs have a larger density than BIm-based DILs with the same EO chains, for example, [DiMPy1O] [TFSI] (1.56 g cm^{-1} at 25℃) has a higher densities than [DiBIm1O] [TFSI] (1.51 g cm^{-1} at 25℃). With the increasing of EO chains, the densities of the corresponding MPy-based DILs and BIm-based DILs are increasing. Moreover, the replacement of the cationic head groups by DES group decreases the fluid density, for example, [DiDES2O] [TFSI] (1.57 g cm^{-1} at 25℃) has a lower density than [DiDES1O] [TFSI] (1.60 g cm^{-1} at 25℃) [26].

4.3.2 Flammability test

A battery might be placed in more hazardous conditions, such as in the presence of fire or sparks. In order to investigate the flammable behavior of the mixed electrolytes containing DIL and [MImEt] [DCA], we also carried out a flammability test called self-extinguishing time (SET) [107]. Each electrolyte was tested three times: the burner was switched on above the sample for 5 s and then switched off. The ignition time after flame setting and the self-extinguish time after removing the burner flame were measured as indices of the non-flammability [108] and the time it took for the flame to extinguish was normalized against liquid mass to give the SET in sg^{-1} [109]. The electrolyte was judged to be nonflammable if the electrolyte never ignited during the testing, or if the ignition of electrolyte ceased when the flame was removed [104].

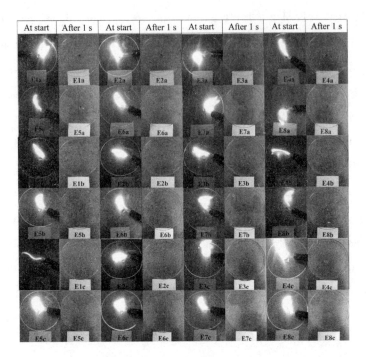

Figure 4.1 Development of the flames for different DILs based electrolytes during SET test directly at the burner was turned on (at start) and was turned off (after 1s) respectively.

It is clearly observed in Figure 4.1 that all the three series of electrolytes with various DILs are certified to be nonflammable, since they cannot be ignited, meaning their SET is all measured to be 0 s. We have also confirmed that neat DILs are completely non-flammable in nature, and contribute to the high safety of the electrolyte system.

4.3.3 Solubility test

The synthesized DILs were tested for their solubility behavior in polar and non-polar solvents (shown in Table 4.2). DILs derived from [DCA] anions exhibited good solubility in polar solvents such as dimethyl sulfoxide (DMSO), acetonitrile (MeCN), methanol (MeOH), water (H_2O), acetone and tetrahydrofuran (THF). Simultaneously, all of them are partially miscible in dichloromethane (DCM) and ethyl acetate (EtOAc), and non-soluble in non-polar solvents such as hexane, diethyl ether (Et_2O). Meanwhile, the solubility behavior of another set of DILs derived from [TFSI] anion display different solubility properties, they are well soluble in DCM and EtOAc, but not soluble in H_2O [50]. So DILs cation replacing the anions with [DCA] increases the solubility of the ionic liquid in water while replacement with the [TFSI] ion dramatically decreases the solubility in water.

Table 4.2. Comparison of solubility in various organic solvents

Compounds	Miscibility									
	DMSO	MeCN	MeOH	H_2O	Et_2O	DCM	EtOAc	Acetone	THF	Hexane
[MImEt] [DCA]	m	m	m	m	i	p	p	m	m	i
[MImEt] [TFSI]	m	m	m	i	i	m	m	m	m	i
[DiMPy1O] [DCA]	m	m	m	m	i	p	p	m	m	i
[DiMPy2O] [DCA]	m	m	m	m	i	p	p	m	m	i
[DiMPy3O] [DCA]	m	m	m	m	i	p	p	m	m	i
[DiMPy4O] [DCA]	m	m	m	m	i	p	p	m	m	i
[DiMPy1O] [TFSI]	m	m	m	i	i	m	m	m	m	i
[DiMPy2O] [TFSI]	m	m	m	i	i	m	m	m	m	i
[DiMPy3O] [TFSI]	m	m	m	i	i	m	m	m	m	i
[DiMPy4O] [TFSI]	m	m	m	i	i	m	m	m	m	i
[DiBIm1O] [DCA]	m	m	m	m	i	p	p	m	m	i
[DiBIm2O] [DCA]	m	m	m	m	i	p	p	m	m	i
[DiBIm3O] [DCA]	m	m	m	m	i	p	p	m	m	i
[DiBIm4O] [DCA]	m	m	m	m	i	p	p	m	m	i
[DiBIm1O] [TFSI]	m	m	m	i	i	m	m	m	m	i
[DiBIm2O] [TFSI]	m	m	m	i	i	m	m	m	m	i
[DiBIm3O] [TFSI]	m	m	m	i	i	m	m	m	m	i

continued

Compounds	Miscibility								
	DMSO	MeCN	MeOH	H$_2$O	Et$_2$O	DCM	EtOAc	Acetone	THF	Hexane
[DiBIm4O][TFSI]	m	m	m	i	i	m	m	m	m	i
[DiDES1O][DCA]	m	m	m	m	i	p	p	m	m	i
[DiDES2O][DCA]	m	m	m	m	i	p	p	m	m	i
[DiDES3O][DCA]	m	m	m	m	i	p	p	m	m	i
[DiDES4O][DCA]	m	m	m	m	i	p	p	m	m	i
[DiDES1O][TFSI]	m	m	m	i	i	m	m	m	m	i
[DiDES2O][TFSI]	m	m	m	i	i	m	m	m	m	i
[DiDES3O][TFSI]	m	m	m	i	i	m	m	m	m	i
[DiDES4O][TFSI]	m	m	m	i	i	m	m	m	m	i

i, *immiscible*; *m*, *miscible*; *p*, *partly miscible*

4.3.4 Ionic Conductivities

All the measurements were taken from 25 ℃ to 70 ℃ via a two-electrode cell and a Gamry potentiostat. The conductivity values were then calibrated to a standard 0.100 mol kg^{-1} KCl aqueous solution with a conductivity of 0.0128 S cm^{-1} at 25.0 ℃ [69].

The electrolytic mixtures were prepared by mixing (in the proper mole fractions) LiTFSI with DILs/[MImEt][DCA], which were continuously stirred until to obtain homogeneous samples (e.g., full dissolution of LiTFSI in ionic liquids) [110]. The ternary mixtures 0.5 M LiTFSI in DILs/[MImEt][DCA] (the DILs:[MImEt][DCA] weight ratio was fixed equal to 1:1) were prepared as electrolytes, with their formulations listed in Table 4.3. The electrolytes were prepared and characterized in a controlled environment dry room.

Table 4.3. Formulations of the electrolytes with 0.5M LiTFSI in DILs/ [MImEt] [DCA]

Formulation	DILs abbreviation	Components weight (g)			Ionic conductivity (mS/cm)	
		DILs	[MImEt] [DCA]	LiTFSI	25℃	70℃
E1a	[DiMPy1O] [DCA]	0.1002	0.1008	0.0246	6.2	22.3
E2a	[DiMPy2O] [DCA]	0.1	0.1	0.0242	9.1	25.3
E3a	[DiMPy3O] [DCA]	0.1015	0.0987	0.0242	5.2	18.54
E4a	[DiMPy4O] [DCA]	0.0995	0.1002	0.0240	3.68	15.19
E5a	[DiMPy1O] [TFSI]	0.1103	0.1105	0.0238	8.1	26.6
E6a	[DiMPy2O] [TFSI]	0.0985	0.1	0.0213	9.74	28.5
E7a	[DiMPy3O] [TFSI]	0.0998	0.102	0.0216	6.95	22.1
E8a	[DiMPy4O] [TFSI]	0.1022	0.1002	0.0215	4.68	16.37
E1b	[DiBIm1O] [DCA]	0.1022	0.1002	0.0245	5.23	21.4
E2b	[DiBIm2O] [DCA]	0.1021	0.0998	0.0244	5.01	18.71
E3b	[DiBIm3O] [DCA]	0.1012	0.1001	0.0243	4.93	17.99
E4b	[DiBIm4O] [DCA]	0.102	0.1002	0.0244	4.77	16.98
E5b	[DiBIm1O] [TFSI]	0.1001	0.1002	0.0219	5.19	18.51
E6b	[DiBIm2O] [TFSI]	0.1048	0.105	0.0228	6.85	20.9
E7b	[DiBIm3O] [TFSI]	0.1032	0.1025	0.0223	6.11	18.9
E8b	[DiBIm4O] [TFSI]	0.1024	0.1006	0.0218	8.35	20.8
E1c	[DiDES1O] [DCA]	0.101	0.1008	0.0241	5.08	19.45
E2c	[DiDES2O] [DCA]	0.1017	0.1001	0.0241	5.05	17.55
E3c	[DiDES3O] [DCA]	0.1007	0.1	0.0240	4.49	17.73
E4c	[DiDES4O] [DCA]	0.1017	0.1001	0.0242	8.19	22.7
E5c	[DiDES1O] [TFSI]	0.1176	0.116	0.0249	6.31	19.99
E6c	[DiDES2O] [TFSI]	0.102	0.0998	0.0216	6.13	19.22
E7c	[DiDES3O] [TFSI]	0.1017	0.1006	0.0219	6.84	20.8
E8c	[DiDES4O] [TFSI]	0.1012	0.1003	0.0219	6.78	24.3

Table 4.4. Formulations of the electrolytes with 0.5M and 0.7M LiTFSI in DILs / [MImEt] [DCA]

Formulation	Components weight (g)				Ionic conductivity (mS/cm)		Notes
	Compounds	DILs	[MImEt] [DCA]	LiTFSI	25℃	70℃	
E2a	[DiMPy2O] [DCA]	0.1	0.1	0.0242	9.1	25.3	0.5M LiTFSI
E6a	[DiMPy2O] [TFSI]	0.0985	0.1	0.0213	9.74	28.5	
E9a	[DiMPy2O] [DCA]	0.1009	0.1018	0.0343	5.79	20	0.7M LiTFSI
E10a	[DiMPy2O] [TFSI]	0.1033	0.104	0.0312	7.44	22.5	

Table 4.5. Formulations of the electrolytes with 0.5M LiTFSI in DILs/ [MImEt] [TFSI]

Formulation	Components weight (g)				Ionic conductivity (mS/cm)		Notes
	Compounds	DILs	[MImEt] [DCA]	LiTFSI	25℃	70℃	
E11a	[DiMPy2O] [DCA]	0.1001	0.1006	0.0217	2.4	11.25	0.5M LiTFSI
E12a	[DiMPy2O] [TFSI]	0.1139	0.114	0.0215	2.8	10.32	

In Table 4.4, the ternary mixtures 0.7 M LiTFSI in [DiMPy2O] [DCA] / [MImEt] [DCA] (E9a) along with [DiMPy2O] [TFSI] (E10a) were prepared, and in Table 4.5, another ternary mixtures 0.5 M LiTFSI in [DiMPy2O] [DCA] / [MImEt] [TFSI] (E11a), along with [DiMPy2O] [TFSI] (E12a) were prepared as parallel groups. And their ionic conductivity data are listed in Figure 4.8.

Many factors can affect their conductivity, such as viscosity, density, ion size, anionic charge delocalization, aggregations and ionic motions. Firstly, it is apparent that ionic conductivity increases with increasing temperature for all the investigated neat DILs and DILs based electrolytes. The decrease in viscosity [111] and faster migration of carrier ions and easier motion in the electrolyte system at higher temperatures are possibly responsible for

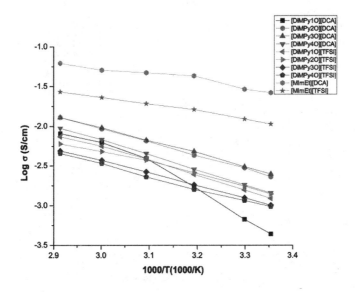

Figure 4.2. Ionic Conductivity versus Temperature of neat MPy-based DILs

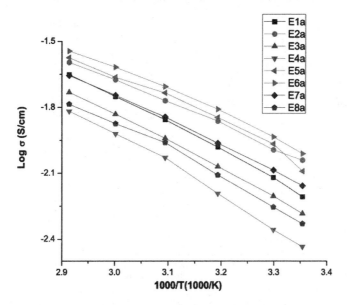

Figure 4.3. Ionic Conductivity versus Temperature of MPy-based electrolytes

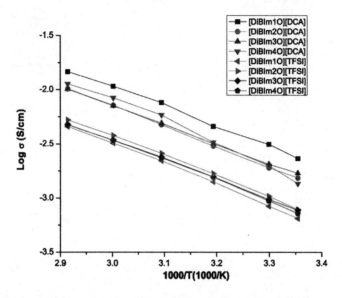

Figure 4.4. Ionic Conductivity versus Temperature of neat BIm-based DILs

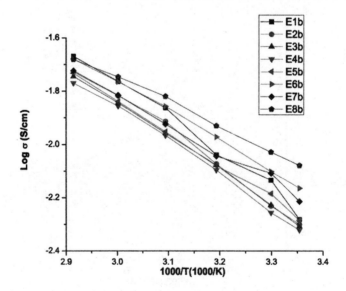

Figure 4.5. Ionic Conductivity versus Temperature of MIm-based electrolytes

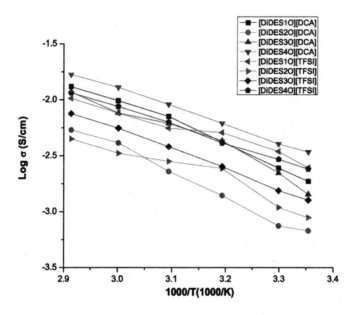

Figure 4. 6. Ionic Conductivity versus Temperature of neat DES-based DILs

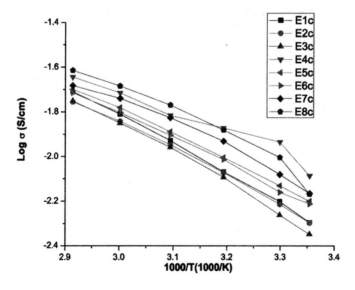

Figure 4. 7. Ionic Conductivity versus Temperature of DES-based electrolytes

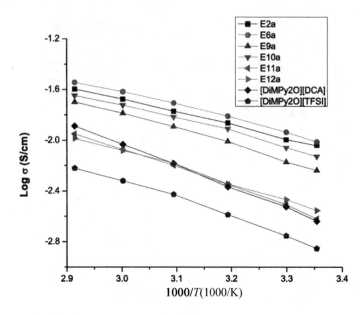

Figure 4.8. Ionic Conductivity versus Temperature of [DiMPy2O] -based electrolytes with different components

this phenomenon [112].

Secondly, the series of neat DILs show pretty poor conductivity. As we know, the conductivity of ILs would decrease significantly upon mixing with a lithium salt [111]. So the lithium-ion batteries fabricated only using neat DILs as the electrolyte may exhibit more disappointing ionic conductivity, which can be ascribed to the following two reasons: (1) the decreasing ionic conductivity of the electrolytes caused by the increased viscosity at low temperature; (2) the largely decreased lithium-ion diffusion rate across electrolyte system at low temperature. To overcome this limitation of DILs, they must be blended with other low viscosity and high stability solvents. Then the ternary electrolyte system, 0.5 M LiTFSI in DILs/ [MImEt] [DCA] (the DILs: [MImEt] [DCA] weight ratio was fixed equal to 1∶1), was developed. [MImEt] [DCA] with pretty low viscosity properties (14.82 cp) acts as a thinner solvent in this electrolyte system. The addition of [MImEt] [DCA] is able to enhance the mobility of those dissociated ions by decreasing the viscosity of the mixture, and hence, increase the conductivity of the whole system, compared with the single solvent electrolyte [113]. Meanwhile, DILs serve not only as non-flammable and non-volatile plasticizers but also as ionic carriers, because DILs with different length of EO chains are good solvents for lithium salts (LiTFSI) forming homogeneous liquid solutions. So compared with neat DILs (Figure 4.2, 4.4, 4.6), the series of DILs based electrolytes (Figure 4.3, 4.5, 4.7) show much higher conductivity.

Thirdly, from Figures 4.3, 4.5, 4.7, it can be seen that E6a exhibits the highest ionic conductivity over the whole temperature range studied, with a conductivity of 9.74 mS cm^{-1} at 25 ℃ (see Table 3). [DiMPy2O] [TFSI] has two EO chains, and one methyl group on each pyrrolidine ring, the closely related [DiMPy2O] [DCA] in E2a also with two EO chains but with different anions [DCA] shows the second highest ionic conductivity (Table 4.3). The ether groups are effective in lowering the crystallinity of DILs and also in improving the ionic conductivity [114]. Thus, ionic conductivity depends on the length of EO chains in the cations and on the concentration of the salt applied [114].

Fourthly, Figure 4.8 suggests that the concentration of charge carriers of these DILs based electrolytes predominate over their segmental mobility in governing the conductivities [48]. By comparing E2a and E9a (see Table 4.4), with increasing the concentration of LiTFSI in the composite electrolyte from 0.5M up to 0.7M, the ionic conductivities decrease from 9.1 mS.cm^{-1} to 5.79 mS.cm^{-1} at 25 ℃. The similar trend also happens between E6a and

E10. This behavior in conductivity is a general feature of DILs based electrolytes, and can be explained as a trade-off between increasing number of charge carriers and reduced chain mobility upon salt addition [115]. While [MImEt][TFSI] instead of [MImEt][DCA] as a thinner solvent, the ionic conductivity of E11a is only 2.4 mS · cm^{-1} and that of E12a is only 2.8 mS · cm^{-1} at 25 ℃ (see Table 4.5).

4.3.5 Cyclic Voltammetry

Figure 4.9 shows the cyclic voltammetry of best four DILs based electrolytes.

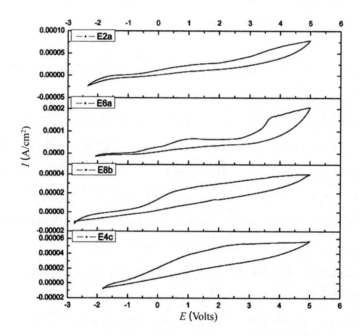

Figure 4.9. CV curves for best four DILs based electrolytes

As is shown, all the four electrolytes exhibit pretty similar high voltage stability. E2a shows the electrochemical stability in a potential range of −2.5 ~ 5.0 V; E6a exhibits that in a range of −2 ~ 3.8 V, E8b performs that in a range of − 2. 5 ~ 4. 5 V and E4c displays that in a range of − 1. 8 ~ 4. 5 V.

E8b has the widest electrochemical window when the anodic potential is swept to the very positive region, an oxidation current arises at + 4. 5 V, suggesting the oxidative decomposition of the ILs system. This onset potential (+ 4. 5 V) can be regarded as a "cathodic limit" of E8b, which defines the highest potential for the DILs-based electrolyte to be stable at. Obviously, this potential value of the cathodic limit is far positive than the highest charging potential (~ 3 V) of the sulfur cathode and therefore it is expected that not only can E8b electrolyte be used for rechargeable sulfur cathodes, but also all the other three electrolyte systems. When the potential is scanned to the negative potential region, the CV curve shows a featureless background line until − 2. 5 V, indicating the chemical and electrochemical stability of the ILs system against lithium. With the presence of lithium ions in the electrolyte, the cathodic peak appeared at 0. 5 V shows the electrochemical deposition of Li ions and the anodic peak appeared at − 2. 5 V indicates successive anodic oxidation of metallic lithium [116]. The appearance of a pair of oxidation-reduction peaks of lithium suggests that lithium can be very well charged and discharged in the E8b electrolyte system without any electrochemical hindrance. Except for these features, there are no any other current bands observed in the CV curve, showing a noble electrochemical stability in the potential range of − 2. 5 ~ 4. 5 V and a sufficient potential window of the 0. 5 M LiTFSI in [DiBIm4O] [TFSI] / [MImEt] [DCA] (E8b) for Li/S cells [117]. The stability of the DILs based electrolytes to achieve high voltages against lithium is confirmed, which ensures their capability for further use in high potential electrolyte materials.

4.4　Summary

In summary, three series of DILs based on MPy, BIm and DES have been successfully synthesized and characterized. All prepared DILs display good thermal stability up to 200 ℃ and non-flammability. Moreover, the ternary electrolytes of 0.5 M LiTFSI in DILs combination with [MImEt][DCA] at 1∶1 weight ratios have been prepared, and their properties have been investigated. The DILs-based electrolytes show wide electrochemical stability and very good ion conductivity. Such impressive results, especially excellent battery performance at low-medium temperatures, make the DILs-based electrolytes prepared in this work promising electrolytes for very safe lithium rechargeable batteries.

Chapter 5 New Asymmetric Fluorinated Room Temperature Ionic Liquids

5.1 Introduction

The current lithium rechargeable battery system has an intrinsic safety problem arising from the use of volatile and flammable organic carbonate as electrolyte solvents. Besides the potential safety issue, the organic carbonate-based electrolyte systems for lithium rechargeable batteries have many other drawbacks, such as low charge-discharge capacity, limitt to operating temperature, and low ionic conductivity, and therefore there is room for possible improvement.

In the past two decades, room temperature ionic liquids (RTILs) have caused a green revolution in chemistry, physics, and materials science in terms of both conceptual innovation and technological breakthroughs. RTILs have also received extensive interest as potential electrolytes for lithium batteries, due to their favorable physical-chemical properties, such as controlled miscibility, non-volatile, non-flammability, and environmental harmlessness [111]. However, although many RTILs have been reported, most are still not suitable for electrochemical applications because of poor conductivity (<1 mS) and/or high viscosity (>100 cP) at room temperature, also due to a poor electrochemical window.

The possibility to modify physical and chemical properties of ILs by tuning the structure through substitution and structural modification of the anions or of the cations [118] allows for the creation of the tailor designed properties. While alternating the ion structure, the reduction of symmetry is known to cause a melting point drop. That is because the anion-cation interaction is minimized due to the asymmetric and delocalized charge. Moreover, as mentioned above, the asymmetric chelation structure could increase the lithium ion transport and ionic conductivity.

On the other hand, the ethylene oxide (EO) unit made by $C-O$, $C-C$, $C-H$ bonds is the best solvating medium [111]. The $-CH_2CH_2O-$ unit provides a suitable space for Li-ion to make chelating with oxygen, and Li-ion migration is associated with the segmental mobility of the EO chain [93].

Meanwhile, by modifying the anion or the fluorine-containing substituent on the cation, a wide range of properties of fluorine-containing ILs can be tuned, including, for example, viscosity, melting point, density, conductivity, solubility, liquid range, thermal and hydrolytic stability, and heat of formation. The influence of inter-ion hydrogen bonding can be diminished by introducing fluorine-containing groups into the cation and especially when employing anions such as bis (trifluorosulfonyl) amide [119]. Among the tools available to synthetic chemists to tune ILs properties, selective fluorination is rather productive, since the introduction of diversely located fluoro substituents including terminal chain, linking arm or core position provides an excellent opportunity for investigating the relationship between structures and properties and for modifying and optimizing the physical/chemical properties of compounds [74]. With the known information that fluorinated substituents in carbonate, molecules help to form stable solid electrolyte interface on the anode surface; fluorinated substituents also suppress the flammability of the molecule; introduction of fluorine into organic molecules usually reduces its boiling temperature and viscosity.

Therefore, the combination of a fluorinated functional group with an EO chain in the cation may result in ILs that display non-volatility, non-flammability, relatively wide electrochemical windows [73] and also low viscosities and high conductivities [72]. Especially it could be a worthy alternative to using-CF_3 or-CF_2CF_2H group incorporated in the structure to decrease the melting point and the viscosity of ILs [43]. Investigations involving this kind of RTILs-bearing fluoro substituents have been reported only rarely [40], in other words, this area is almost an unexploited subject.

On the other hand, DCA anions offer exceptionally low viscosity, which is of benefit for heat and mass transport to provide high conductivities. While being relatively delocalized, still retains some Lewis basicity, DCA anion based ILs have highly desirable characteristics for potential Li electrolyte solvents. In the meantime, DCA ILs can be manufactured relatively inexpensively because DCA is already a commodity anion, containing only carbon and nitrogen, which is produced in large amounts for the pharmaceutical industry [119]. However, for lithium batteries, the DCA anodic stability is below that of TFSI. The TFSI anion is an exceedingly weak base as an effect of strong delocalization of charge throughout the molecule. This feature is contributing to its stability vs. oxidation. The delocalized charge, weak basicity, and poor hydrogen-bonding ability [120], producing weaker ion-ion interactions, coupled with the flexibility of the molecule tend to produce comparatively low viscosity and good thermal stability ILs [121]. Both the TFSI and DCA anions can facilitate stable charge/discharge of the Li-S cells.

In this work, we have described the design, synthesis, and characterization of twelve novel unsymmetrical fluorinated cationic-based ILs with an EO unit. This series of novel ILs is named as fluorinated room temperature ionic liquids (FRTILs). Their properties and potential applications as solvents for the electrolytes, high ion transport materials have also been extensively developed.

The understanding of the molecular interactions of ILs in an electrolyte system is a prerequisite for sustainable predicting, controlling, and designing their co-solvent properties and their application in an industrial process.

5.2 Experimental

5.2.1 Materials

2, 2, 2 - Trifluoroethanol and 2, 2, 3, 3 - Tetrafluoro - 1 - propanol were purchased from SynQuest Labs, Inc. Sodium hydroxide, Ethylene carbonate, 4 - Toluenesulfonyl chloride, Lithium bis (trifluoromethyl sulfonyl) imide, 1 - methylpyrrolidine, 1 - butyl imidazolium, and diethyl sulfide were purchased from Sigma-Aldrich Chemical Co., Ltd (USA). All the solvents were supplied by Fisher Scientific Co., Ltd., (USA).

5.2.2 General procedures for the preparation of FRTILs

We reported the synthesis, characterization and thermal properties of twelve new FRTILs. The general molecular structures and abbreviations of these compounds are listed in Table 5.1.

Table 5.1　Appendix abbreviations of FRTILs

Cations \ Anions	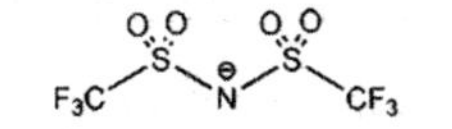 F₃C-S(O)₂-N⁻-S(O)₂-CF₃	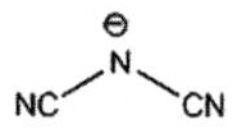 NC-N⁻-CN
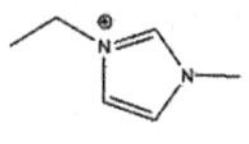 Ethyl-methylimidazolium	[MImEt] [TFSI]	[MImEt] [DCA]
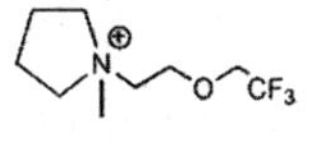 Methylpyrrolidinium-OCH₂CH₂OCF₃	[MPyOTF] [TFSI]	[MPyOTF] [DCA]
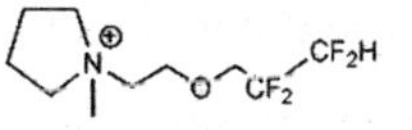 Methylpyrrolidinium-OCH₂CH₂OCF₂CF₂H	[MPyOPF] [TFSI]	[MPyOPF] [DCA]
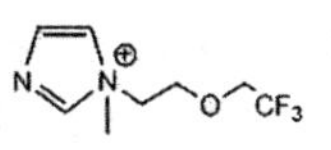 Methylimidazolium-OCH₂CH₂OCF₃	[MImOTF] [TFSI]	[MImOTF] [DCA]
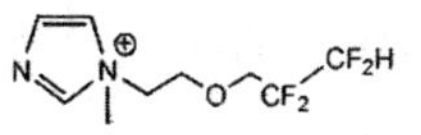 Methylimidazolium-OCH₂CH₂OCF₂CF₂H	[MImOPF] [TFSI]	[MImOPF] [DCA]
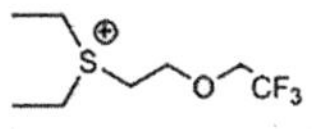 Diethylsulfonium-OCH₂CH₂OCF₃	[DESOTF] [TFSI]	[DESOTF] [DCA]
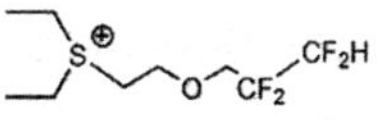 Diethylsulfonium-OCH₂CH₂OCF₂CF₂H	[DESOPF] [TFSI]	[DESOPF] [DCA]

5.2.2.1　General procedures for the preparation of MPy-based FRTILs.

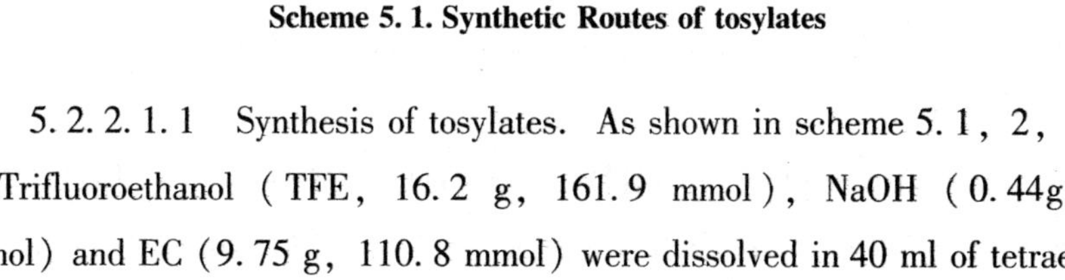

Scheme 5.1. Synthetic Routes of tosylates

5.2.2.1.1　Synthesis of tosylates. As shown in scheme 5.1, 2, 2, 2 - Trifluoroethanol (TFE, 16.2 g, 161.9 mmol), NaOH (0.44g, 11 mmol) and EC (9.75 g, 110.8 mmol) were dissolved in 40 ml of tetraethylene glycol dimethyl ether (G4) and refluxed at 150 ℃ for 4h. Then the product was distilled under vacuum. The distillate at a range of 112 ~ 150 ℃ was collected. A colorless liquid product 2 - (2, 2, 2 - trifluoroethoxy) ethanol (TFEE, yield: 14.53 g, 90%) was prepared. Then TFEE (14.53 g, 100.84 mmol), and TsCl (21.10 g, 111.1 mmol) were dissolved in 100 ml of DCM. NaOH (8.07 g, 201.8 mmol) was added slowly in the solution with an ice bath under vigorous stirring. After stirring overnight at room temperature, to which another 50 ml of DCM was added to dilute the mixture, and then it was washed with 50 ml of water one time and 20 ml of brine two times. The organic layer was separated and dried with anhydrous $MgSO_4$. Then the solvent was removed to get the desired product: 2 - (2, 2, 2 - trifluoroethoxy) ethyl 4 - methylbenzenesulfonate (TFEEOTs, yield: 27.54 g, 91.6%), which was a colorless viscous liquid and crystallized at room temperature.

An analogous procedure was used to prepare 2 - (2, 2, 3, 3 - tetrafluoropropoxy) ethanol (TFPE) started with 2, 2, 3, 3 - Tetrafluoro - 1 - propanol (TFP, 21.7 g, 164.3 mmol), NaOH (0.43 g, 10.8 mmol) and EC (9.65 g, 109.6 mmol), to give the desired compound as a colorless liquid: 2 - (2, 2, 3, 3 - tetrafluoropropoxy) ethanol (TFPE, yield: 16.57 g, 85.8%). The corresponding tosylate was prepared analogously from TFPE (16.57 g, 94.09 mmol) and TsCl (19.7 g, 103.7 mmol), NaOH (7.52 g, 188 mmol) to give the desired compound 2 - (2, 2, 3, 3 - tetrafluoropropoxy) ethyl 4 - methylbenzenesulfonate (TFPEOTs, yield: 85.98 g, 91.4%), which was a colorless viscous liquid and crystallized at room temperature.

5.2.2.1.2 Synthesis of FRTILs [MPyOTF] [TFSI]. TFEEOTs (8.0 g, 26.8 mmol) and 1 - methylpyrrolidine (4.58 g, 53.8 mmol) were mixed

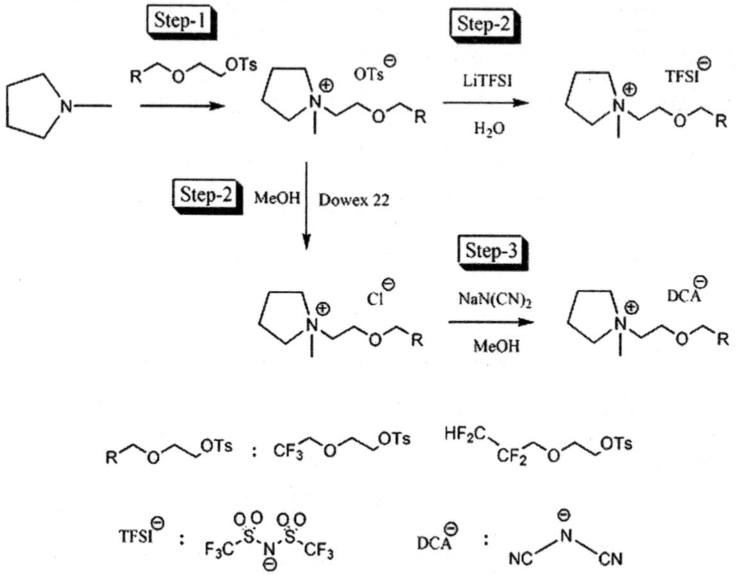

Scheme 5. 2. Synthetic Routes of MPy-based FRTILs

and refluxed overnight under stirring. The mixture was cooled down to room temperature. As ILs are insoluble in ethyl ether, it develops an insoluble sticky mass inside the flask. We can easily decant ether avoiding compound loss where starting materials will go away with ether. Further washing the sticky mass twice with ether [122] and removing the solvent gave the pure desired compound [MPyOTF] OTs (yield: 7.52 g, 73.1%) as a brown viscous liquid.

[MPyOTF] OTs (2.87 g, 7.5 mmol) with LiTFSI (2.38 g, 8.3 mmol) was dissolved in 15 ml of deionized water. The solution was stirred overnight at room temperature, the solution was extracted with 30 ml of dichloromethane. The obtained organic layer was washed three times with water of 10 ml, dried over anhydrous magnesium sulfate and active carbon. The residue was purified by column chromatography on alumina and concentrated under reduced pressure to give the final product [MPyOTF] [TFSI] (yield: 3.18 g, 86.4%) as a brown liquid.

^{1}H NMR (300 MHz, D-acetone): δ (ppm) 4.24 (t, J = 5.1 Hz, 2H), 4.13 (q, J = 7.8 Hz, 2H), 3.85 (t, J = 4.2 Hz, 2H), 3.75 (t, J = 3.51 Hz, 4H), 3.29 (s, 3H), 2.39 - 2.05 (m, 4H). ^{13}C NMR (300 MHz, CDCl$_3$): δ (ppm) 122.8, 119.8 (q, J = 159 Hz), 68.33, 66.2, 65.3, 62, 47.93, 21.2.

5.2.2.3 Synthesis of FRTILs [MPyOPF] [TFSI]. An analogous procedure was used to prepare [MPyOPF] OTs started with TFPEOTs (7.74 g, 23.4 mmol), 1-methylpyrrolidine (3.98 g, 46.7 mmol), to give the desired compound as a dark brown liquid [MPyOPF] OTs (yield: 7.62 g, 18.34 mmol). The corresponding ionic liquid was prepared analogously from [MPyOPF] OTs (2.25 g, 5.4 mmol) and LiTFSI (1.85 g, 6.4 mmol) to give the desired compound as a dark brown liquid [MPyOPF] [TFSI] (yield: 2.34 g, 82.4%);

^{1}H NMR (300 MHz, D-acetone): δ (ppm) 6.34 (t, J = 5.4 Hz, 1H), 4.22 (t, J = 3.51 Hz, 2H), 4.09 (t, J = 5.1 Hz, 2H), 3.87 (t, J = 4.5 Hz, 2H), 3.59 (t, J = 3.51 Hz, 4H), 3.29 (s, 3H), 2.35 - 2.04 (m, 4H). ^{13}C NMR (300 MHz, D-acetone): δ (ppm) 128.0, 119.8 (q, J = 159 Hz), 109.5, 70, 67.6, 57.8, 49.4, 35.7, 20.7.

5.2.2.4 Synthesis of FRTILs [MPyOTF] [DCA]. [MPyOTF] OTs (3.76 g, 9.8 mmol) was dissolved in 15 ml of methanol and passed through the column with 25 ml dried Dowex © 22 resin at the dropping rate of 1 drop/second, to which 60mL methanol was added to wash the column. Then all methanol solutions were combined, and the solvent was removed. The desired compound [MPyOTF] Cl (yield: 2.07g, 85.1%) was obtained as a brown viscous liquid.

NaN(CN)$_2$ (0.34 g, 3.8 mmol) was totally dissolved in 30ml of methanol, to which [MPyOTF] Cl (2.07 g, 8.4 mmol) in another 5 ml of methanol was added together. The mixture was stirred overnight at room temperature, methanol was removed by vacuum distillation until some precipitation came out. Then the precipitation was filtered away, to which 15 ml of acetone was added to the filtrate. Removed the solvent by vacuum distillation until some precipitation came out again. Repeated the foregoing procedures for several times until no precipitation came out. Then the filtrate was concentrated under reduced pressure to give the final product [MPyOTF] [DCA] (yield: 1.63 g, 70.1%) as a dark brown liquid.

^{1}H NMR (300 MHz, D$_2$O): δ (ppm) 4.1 (t, J = 5.1 Hz, 2H), 3.73 (q, J = 7.8 Hz, 2H), 3.38 (t, J = 4.2 Hz, 2H), 3.18 (t, J = 3.51 Hz, 4H), 2.87 (s, 3H), 2.37 - 1.91 (m, 4H). ^{13}C NMR (300 MHz, D$_2$O): δ (ppm) 119.9, 65.9, 55.8, 51.7, 48.4, 40.6, 21.2.

5.2.2.5 Synthesis of DILs [MPyOPF] [DCA]. An analogous procedure was used to start with [MPyOPF] OTs (1.91 g, 4.6 mmol) to give the desired compound as a yellow liquid [MPyOPF] Cl (yield: 1.08 g, 83.6%). The corresponding ionic liquid was prepared analogously from [MPyOPF] Cl (1.08, 3.9 mmol) and NaN(CN)$_2$ (0.72 g, 8.1 mmol) to give the desired compound as a dark brown liquid [MPyOPF] [DCA] (yield: 0.86 g, 71.8%).

^{1}H NMR (300 MHz, D$_2$O): δ (ppm) 6.15 (t, J = 5.4 Hz, 1H), 4.05 (t, J = 3.51 Hz, 2H), 3.59 (t, J = 5.1 Hz, 2H), 3.51 (t, J = 4.5 Hz, 2H), 3.07 (t, J = 3.51 Hz, 4H), 3.03 (s, 3H), 2.16 - 1.96 (m, 4H). ^{13}C NMR (300 MHz, D$_2$O): δ (ppm) 128.0, 119.9, 109.5, 67.2, 63.0, 58.1, 55.8, 48.4, 21.1.

5.2.3 General procedures for the preparation of MIm-based FR-TILs

5.2.3.1 Synthesis of FRTILs [MImOTF][TFSI]. TFEEOTs (7.41 g, 24.9 mmol) and 1-methylimidazole (4.16 g, 48.9 mmol) were mixed and refluxed overnight under stirring. The mixture was cooled down to room temperature. After the solvent had been removed by vacuum evaporation, the residue was washed three times with 15 ml of ethyl ether. The remaining volatile was removed under high vacuum to give the desired compound [MImOTF]OTs (yield: 7.38 g, 78.1%) as a yellow viscous liquid.

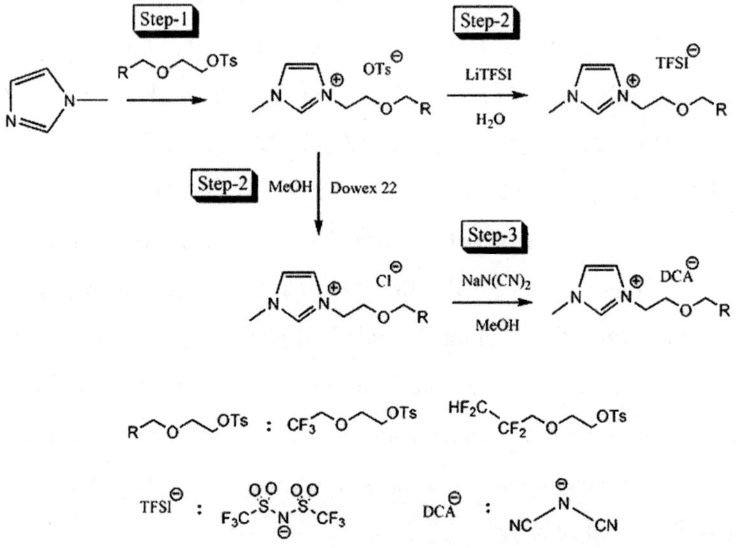

Scheme 5.3. Synthetic Routes of MIm-based FRTILs

[MImOTF]OTs (2.46 g, 6.5 mmol) with LiTFSI (2.04 g, 7.1 mmol) was dissolved in 15 ml of deionized water. The solution was stirred overnight at room temperature, the solution was extracted with 30 ml dichloromethane. The obtained organic layer was washed three times with water of 10 ml, dried over anhydrous magnesium sulfate and active carbon. The residue was purified by column chromatography on alumina and concentrated under reduced pressure to give the final product [MImOTF] [TFSI] (yield: 2.16 g, 68.3%) as a yellow liquid.

^{1}H NMR (300 MHz, D-acetone): δ (ppm) 9.05 (s, 1H), 7.85 (d, J = 1.2 Hz, 1H), 7.72 (d, J = 1.2 Hz, 1H), 4.60 (t, J = 4.5 Hz, 2H), 4.12 (m, 2H), 4.05 (t, J = 6.9 Hz, 2H), 3.71 (s, 3H). ^{13}C NMR (300 MHz, D-acetone): δ (ppm) 138, 131, 122.8, 119.8 (q, J = 159 Hz), 70.1, 67, 49.4, 37.

5.2.3.2 Synthesis of FRTILs [MImOPF] [TFSI]. An analogous procedure was used to start with TFPEOTs (7.12 g, 21.6 mmol), 1 - methylimidazole (3.57 g, 41.9 mmol) to give the desired compound as a pale yellow liquid [MImOPF] OTs (yield: 7.72 g, 86.8%). The corresponding ionic liquid was prepared analogously from [MImOPF] OTs (2.46 g, 6.5 mmol) and LiTFSI (2.04 g, 7.1 mmol) to give the desired compound as a pale yellow liquid [MImOPF] [TFSI] (yield: 2.16 g, 68.3%);

^{1}H NMR (300 MHz, D-acetone): δ (ppm) 9.05 (s, 1H), 7.85 (d, J = 1.2 Hz, 1H), 7.72 (d, J = 1.2 Hz, 1H), 6.27 (t, J = 4.8 Hz, 1H), 4.57 (t, J = 4.5 Hz, 2H), 4.11 (t, J = 7.2 Hz, 2H), 3.98 (t, J = 6.9 Hz, 2H), 3.71 (s, 3H). ^{13}C NMR (300 MHz, D-acetone): δ (ppm) 138, 131, 119.8 (q, J = 159 Hz), 70.1, 67, 49.4, 37.

5.2.3.3 Synthesis of FRTILs [MImOTF] [DCA]. [MImOTF] OTs (3.49 g, 9.2 mmol) was dissolved in 15 ml of methanol and passed through the column with 25 ml dried Dowex © 22 resin at the dropping rate of 1 drop/second, to which 60mL methanol was added to wash the column. Then all methanol solutions were combined, and the solvent was removed. The desired compound [MImOTF] Cl (yield: 2.09 g, 93.1%) was obtained as a brown viscous liquid.

NaN(CN)$_2$ (0.84 g, 9.4 mmol) was totally dissolved in 30ml of methanol, to which [MImOTF] Cl (2.09 g, 8.6 mmol) in another 5 ml methanol was added together. The mixture was stirred overnight at room temperature, methanol was removed by vacuum distillation until some precipitation came out. Then the precipitation was filtered away, and 15 ml acetone was added to the filtrate. Removed the solvent by vacuum distillation until some precipitation came out again. Repeated the foregoing procedures for several times until no precipitation came out. Then the filtrate was concentrated under reduced pressure to give the final product [MImOTF] [DCA] (yield: 1.66 g, 70.7%) as a brown liquid.

^{1}H NMR (300 MHz, D$_2$O): δ (ppm) 7.54 (s, 1H), 6.95 (d, J = 1.2 Hz, 1H), 6.87 (d, J = 1.2 Hz, 1H), 5.03 (t, J = 4.5 Hz, 2H), 3.93 (m, 2H), 3.80 (t, J = 6.9 Hz, 2H), 3.61 (s, 3H). ^{13}C NMR (300 MHz, D$_2$O): δ (ppm) 138, 131, 122.8, 121.5, 114.44, 70.0, 49.1, 36.1, 33.2.

5.2.3.4 Synthesis of DILs [MImOPF] [DCA]. An analogous procedure was used to start with [MImOPF] OTs (3.61 g, 8.8 mmol) to give the desired compound as a brown liquid [MImOPF] Cl (yield: 2.23 g, 92%). The corresponding ionic liquid was prepared analogously from [MImOPF] Cl (2.23, 8.1 mmol) and NaN(CN)$_2$ (0.79 g, 8.9 mmol) to give the desired compound as a brown liquid [MImOPF] [DCA] (yield: 1.73 g, 70.1%);

^{1}H NMR (300 MHz, D$_2$O): δ (ppm) 7.54 (s, 1H), 7.05 (d, J = 1.2 Hz, 1H), 6.95 (d, J = 1.2 Hz, 1H), 6.12 (t, J = 4.8 Hz, 1H), 4.68 (t, J = 4.5 Hz, 2H), 4.4 (t, J = 7.2 Hz, 2H), 3.97 (t, J = 6.9 Hz, 2H), 3.86 (s, 3H). ^{13}C NMR (300 MHz, D$_2$O): δ (ppm) 137, 124.7, 123.6, 122.7, 119.7, 109.5, 70.0, 67.3, 48.8, 35.8.

5.2.4 General procedures for the preparation of DES-based FRTILs

5.2.4.1 Synthesis of FRTILs [DESOTF][TFSI]. TFEEOTs (11.8 g, 39.6 mmol) and diethyl sulfide (7.14 g, 83.9 mmol) were directly mixed and refluxed at 70 ℃ under stirring for 3 days, then cooled down to room temperature. The residue was washed three times with 15 ml of ethyl ether. The remaining volatile was removed under high vacuum to give the desired compound [DESOTF] OTs (yield: 4.8 g, 32.6%) as a pale yellow liquid.

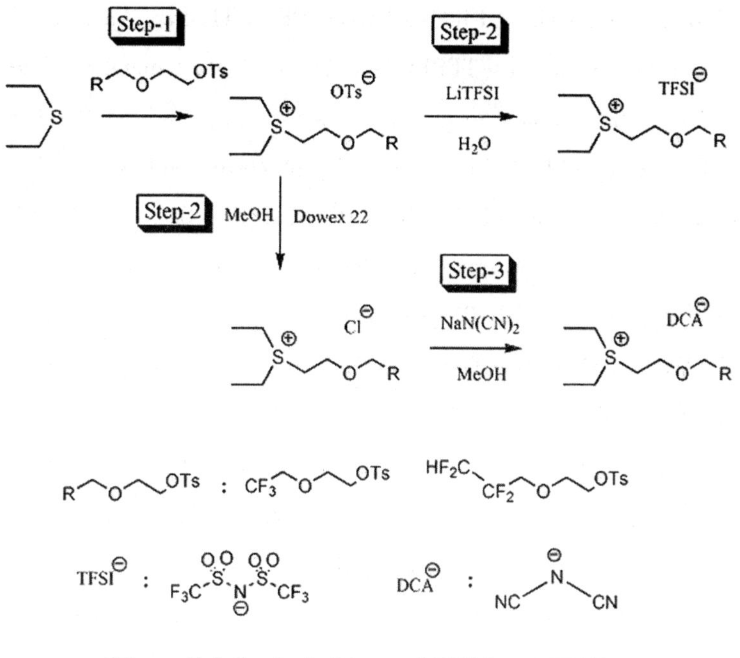

Scheme 5.4. Synthetic Routes of DES-based FRTILs

[DESOTF] OTs (0.83g, 2.2 mmol) with LiTFSI (0.85 g, 3 mmol) was dissolved in 15 ml deionized water. The solution was stirred overnight at room temperature, the solution was extracted with 30 ml dichloromethane.

The obtained organic layer was washed three times with water of 10 ml, dried over anhydrous magnesium sulfate and active carbon. The residue was purified by column chromatography on alumina and concentrated under reduced pressure to give the final product [DESOTF][TFSI] (yield: 0.57 g, 51.3%) as a pale yellow liquid.

^{1}H NMR (300 MHz, D-acetone): δ (ppm) 4.65 (m, 2H), 4.25 (t, J=5.1 Hz, 2H), 3.85 (t, J=5.3 Hz, 2H), 3.7–3.4 (m, 4H), 1.55 (t, J=6.6 Hz, 6H). ^{13}C NMR (300 MHz, D-acetone): δ (ppm) 129.5, 119.8 (q, J=159 Hz), 69.7, 66.4, 33.8, 30.3, 8.0.

5.2.4.2 Synthesis of FRTILs [DESOPF][TFSI]. An analogous procedure was used to start with TFPEOTs (12.59 g, 38.1 mmol), diethyl sulfide (7.2 g, 84.6 mmol) to give the desired compound as a yellow liquid [DESOPF] OTs (yield: 4.48 g, 28%). The corresponding ionic liquid was prepared analogously from [DESOPF] OTs (2.61 g, 6.2 mmol) and LiTFSI (2.14 g, 7.4 mmol) to give the desired compound as a yellow liquid [DESOPF][TFSI] (yield: 2.15 g, 65.5%);

^{1}H NMR (300 MHz, D-acetone): δ (ppm) 6.3 (t, J=1.2 Hz, 1H), 4.25 (t, J=4.8 Hz, 2H), 3.8 (t, J=5.1 Hz, 2H), 3.65 (t, J=5.3 Hz, 2H), 3.5–3.3 (m, 4H), 1.55 (t, J=6.6 Hz, 6H). ^{13}C NMR (300 MHz, D-acetone): δ (ppm) 127.8, 125.4, 119.8 (q, J=159 Hz), 69.7, 66.3, 34.0, 32.5, 8.2.

5.2.4.3 Synthesis of FRTILs [DESOTF][DCA]. [DESOTF] OTs (2.31 g, 6.2 mmol) was dissolved in 15 ml of methanol and passed through the column with 25 ml dried Dowex © 22 resin at the dropping rate of 1 drop/second, to which 60mL methanol was added to wash the column. Then all methanol solutions were combined, and the solvent was removed. The desired compound [DESOTF] Cl (yield: 1.15g, 73.4%) was obtained as a dark yellow liquid.

NaN(CN)$_2$ (0.44 g, 5.0 mmol) was totally dissolved in 15ml of methanol, to which [DESOTF] Cl (1.15 g, 4.5 mmol) in another 5 ml methanol was added together. The mixture was stirred overnight at room temperature, methanol was removed by vacuum distillation until some precipitation came out. Then the precipitation was filtered away, and 15 ml of acetone was added to the filtrate. Removed the solvent by vacuum distillation until some precipitation came out again. Repeated the foregoing procedures for several times until no precipitation came out. Then the filtrate was concentrated under reduced pressure to give the final product [DESOTF] [DCA] (yield: 0.77 g, 60.3%) as a dark yellow liquid.

^{1}H NMR (300 MHz, D$_2$O): δ (ppm) 3.89 (m, 2H), 3.6 (t, J = 5.1 Hz, 2H), 3.35 (t, J = 5.3 Hz, 2H), 3.24 - 3.18 (m, 4H), 1.35 (t, J = 6.6 Hz, 6H). ^{13}C NMR (300 MHz, D$_2$O): δ (ppm) 129.5, 119.9, 69.7, 66.4, 33.8, 30.3, 8.07

5.2.4.4 Synthesis of DILs [DESOPF] [DCA]. An analogous procedure was used to start with [DESOPF] OTs (2.01 g, 4.8 mmol) to give the desired compound as a yellow liquid [DESOPF] Cl (yield: 0.96 g, 70.4%). The corresponding ionic liquid was prepared analogously from [DESOPF] Cl (0.96 g, 3.4 mmol) and NaN(CN)$_2$ (0.33 g, 3.7 mmol) to give the desired compound as a yellow liquid [DESOPF] [DCA] (yield: 0.66 g, 62.1%).

^{1}H NMR (300 MHz, D$_2$O): δ (ppm) 6.1 (t, J = 1.2 Hz, 1H), 4.03 (t, J = 4.8 Hz, 2H), 3.6 (t, J = 5.1 Hz, 2H), 3.35 (t, J = 5.3 Hz, 2H), 3.3 - 3.1 (m, 4H), 1.35 (t, J = 6.6 Hz, 6H). ^{13}C NMR (300 MHz, D$_2$O): δ (ppm) 129.4, 125.4, 119.9, 69.7, 66.3, 38.3, 33.6, 7.9.

5.3 Results and Discussion

5.3.1 Characterization

In order to make the best choice of the electrolyte composition for use in a practical Li rechargeable battery, we should take into account the flammability, conductivity, and viscosity of the electrolytes. All the physicochemical properties of the FRTILs are listed in Table 5.2.

Table 5.2. Physicochemical properties of the FRTILs

Compounds	Molecular weight (g/mol)	Density (g/ml)	Viscosity (cp)	Appearance	Ionic conductivity Neat (mS/cm) 25 ℃	70 ℃
[MPyOTF][DCA]	278.27	1.33	60.6	brown liquid	1.228	9.24
[MPyOPF][DCA]	310.29	1.35	114.9	dark brown liquid	1.504	9.95
[MPyOTF][TFSI]	492.38	1.55	78.94	brown liquid	2.28	10.04
[MPyOPF][TFSI]	524.4	1.59	98.71	dark brown liquid	1.43	7.9
[MImOTF][DCA]	274.34	1.30	17.32	brown liquid	3.55	21.4
[MImOPF][DCA]	306.26	1.35	105	brown liquid	3.65	20.9
[MImOTF][TFSI]	488.35	1.59	49.82	yellow liquid	5.3	16.29
[MImOPF][TFSI]	520.36	1.63	83.59	pale yellow liquid	1.998	9.91
[DESOTF][DCA]	283.31	1.21	27.59	dark yellow liquid	5.19	21.6
[DESOPF][DCA]	315.33	1.32	67.7	yellow liquid	1.046	6.4
[DESOTF][TFSI]	497.42	1.30	38.95	pale yellow liquid	2.2	7.31
[DESOPF][TFSI]	529.44	1.41	44.67	yellow liquid	1.921	6.18

The density is the most often measured and reported physical property of ionic liquids because nearly every application requires knowledge of the density. Table 5.2 contains density values of ILs, in general, ILs are denser than either organic solvents or water. The density properties follow quite consistent trends. For a given cation, the density increases as the molecular weight of the anion increases, just as FRTILs containing TFSI anions have a higher density than those containing DCA anions. For instance, [MPyOTF] [TFSI] (1.55 g cm^{-1} at 25℃) has higher density than [MPyOTF] [DCA] (1.330 g cm^{-1} at 25 ℃), [MImOTF] [TFSI] (1.59 g cm^{-1} at 25℃) has higher density than [MImOTF] [DCA] (1.30 g cm^{-1} at 25℃), [DESOTF] [TFSI] (1.38 g cm^{-1} at 25℃) has higher density than [DESOTF] [DCA] (1.21 g cm^{-1} at 25℃). Since none of the anions have a particularly long or awkward chain, it is possible that they all can occupy favorable, relatively close-approach positions around the cation, resulting in the high densities [106]. On the other hand, MPy-based FRTILs have a larger density than MIm-based FRTILs, and DES-based FRTILs have the lowest density [26].

5.3.2 Flammability test

A battery might be placed in more hazardous conditions, for instance, in the presence of sparks or fire. In order to investigate the flammable behavior of the mixed electrolytes containing FRTIL and [MImEt] [DCA], we carried out a flammability test called self-extinguishing time (SET) [107]. Each electrolyte was tested three times: the burner was switched on above the sample for 5 s and then switched off. The ignition time after flame setting and the self-extinguish time after removing the burner flame were measured as indices of the non-flammability [108] and the time it took for the flame to extinguish was normalized against liquid mass to give the SET in sg^{-1} [109]. The electrolyte was judged to be nonflammable if the electrolyte never ignited during the testing, or if the ignition of electrolyte ceased when the flame was removed [104].

基于新型可充电锂电池固态和液态电解质的研究 >>>

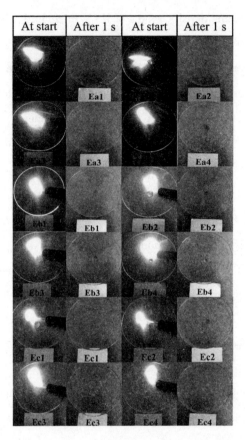

Figure 5.1 Development of the flames for different FRTILs based electrolytes during SET test directly at the burner was turned on (at start) and was turned off (after 1s) respectively.

It is clearly observed from Figure 5.1 that all the three series of electrolytes with various of FRTILs are certified to be nonflammable, since they cannot be ignited, meaning their SET is all measured to be 0 s. We have also confirmed that neat FRTILs are completely non-flammable in nature, and contribute to the high safety of the electrolyte system.

5.3.3 Solubility test

The synthesized FRTILs were tested for their solubility behavior in polar and non-polar solvents (shown in Table 5.3). FRTILs derived from [DCA] anions exhibit good solubility in polar solvents such as dimethyl sulfoxide (DMSO), acetonitrile (MeCN), methanol (MeOH), water (H_2O), acetone and tetrahydrofuran (THF). Simultaneously, all of them are partially miscible in dichloromethane (DCM) and ethyl acetate (EtOAc), and non-soluble in non-polar solvents such as hexane, diethyl ether (Et_2O). Meanwhile, the solubility behavior of another set of FRTILs derived from [TFSI] anion displays different solubility properties, they are well soluble in DCM and EtOAc, but not soluble in H_2O [50]. So FRTILs cation replacing the anions with [DCA] increase the solubility of the ionic liquid in water while replacement with the [TFSI] anions dramatically decreases the water solubility. Meanwhile, the hydrophobic [TFSI] anions based FRTILs have certain additional special properties, including: 1) inertness to organic solvents and oxidizing agents; 2) resistance to corrosive acids and bases; and 3) resistance to extremes of temperature and pressure [119].

Table 5.3. Comparison of solubility in various organic solvents

Compounds	Miscibility									
	DMSO	MeCN	MeOH	H_2O	Et_2O	DCM	EtOAc	Acetone	THF	Hexane
[MImEt] [DCA]	m	m	m	m	i	p	p	m	m	i
[MImEt] [TFSI]	m	m	m	i	i	m	m	m	m	i
[MPyOTF] [DCA]	m	m	m	m	i	p	p	m	m	i
[MPyOPF] [DCA]	m	m	m	m	i	p	p	m	m	i
[MPyOTF] [TFSI]	m	m	m	i	i	m	m	m	m	i
[MPyOPF] [TFSI]	m	m	m	i	i	m	m	m	m	i
[MImOTF] [DCA]	m	m	m	m	i	p	p	m	m	i
[MImOPF] [DCA]	m	m	m	m	i	p	p	m	m	i

continued

Compounds	Miscibility								
	DMSO	MeCN	MeOH	H$_2$O	Et$_2$O	DCM	EtOAc	Acetone	THF	Hexane
[MImOTF][TFSI]	m	m	m	i	i	m	m	m	m	i
[MImOPF][TFSI]	m	m	m	i	i	m	m	m	m	i
[DESOTF][DCA]	m	m	m	m	i	p	p	m	m	i
[DESOPF][DCA]	m	m	m	m	i	p	p	m	m	i
[DESOTF][TFSI]	m	m	m	i	i	m	m	m	m	i
[DESOPF][TFSI]	m	m	m	i	i	m	m	m	m	i

i, immiscible; m, miscible; p, partly miscible

5.3.4 Ionic Conductivities

All measurements were taken and characterized in a controlled environment dry room from 25 ℃ to 70 ℃ via a two-electrode cell and a Gamry potentiostat. The conductivity values were then calibrated to a standard 0.1 mol kg^{-1} KCl aqueous solution with a conductivity of 0.0128 S cm^{-1} [69] at 25.0℃.

As we know the conductivity of ILs would decrease significantly upon mixing with a lithium salt [111]. So the lithium-ion batteries fabricated only using neat FRTILs as the electrolyte may exhibit more disappointing ionic conductivity, which can be ascribed to the following two reasons: (1) the decreasing ionic conductivity of the electrolytes caused by the increased viscosity at low temperature; (2) the largely decreased lithium-ion diffusion rate across electrolyte system at low temperature. To overcome this limitation of FRTILs, they must be blended with other low viscosity and high stability solvents, then the ternary electrolyte system, 0.5 M LiTFSI in FRTILs/[MImEt][DCA] (the FRTILs: [MImEt][DCA] weight ratio was fixed equal to 1:1), was developed. [MImEt][DCA] with pretty low viscosity properties (14.82 cp at 25 ℃) and high ionic conductivity (27 S/cm at 25 ℃) acts as a thinner solvent in this electrolyte system. The addition of [MImEt][DCA] is able to enhance the mobility of those dissociated ions by decreasing the viscosity of the mixture, and hence, increase the conductivity of the whole system, compared with the single solvent electrolyte [113].

The electrolytic mixtures were prepared by mixing (in the proper mole fractions) LiTFSI with FRTILs/ [MImEt] [DCA], which were continuously stirred until to obtain homogeneous samples (e. g., full dissolution of LiTFSI in ionic liquids) [110]. The ternary mixtures, 0.5 M LiTFSI in FRTILs/ [MImEt] [DCA] (the FRTILs: [MImEt] [DCA] weight ratio was fixed equal to 1 : 1), were prepared as electrolytes, and their formulations are listed in Table 5.4. The ionic conductivity data of neat FRTILs are compared in Figures 5.2, 5.4 & 5.6, and the ionic conductivity data of FRTILs based electrolytes are paralleled in Figures 5.3, 5.5 & 5.7.

Table 5. 4. Formulations of the electrolytes with 0.5M LiTFSI in FRTILs/ [MImEt] [DCA]

Formulation	DILs abbreviation	Components weight (g)			Ionic conductivity (mS/cm)	
		FRTILs	[MImEt] [DCA]	LiTFSI	25 ℃	70 ℃
Ea1	[MPyOTF] [DCA]	0.1002	0.1007	0.0233	9.95	29.6
Ea2	[MPyOPF] [DCA]	0.1016	0.1003	0.0232	8.32	23.9
Ea3	[MPyOTF] [TFSI]	0.0999	0.1001	0.0216	9.74	24.6
Ea4	[MPyOPF] [TFSI]	0.1005	0.1	0.0214	8.05	25.3
Eb1	[MImOTF] [DCA]	0.1014	0.1007	0.0236	9.1	30.8
Eb2	[MImOPF] [DCA]	0.1005	0.1006	0.0231	8.29	28
Eb3	[MImOTF] [TFSI]	0.1007	0.1005	0.0215	10.51	29.3
Eb4	[MImOPF] [TFSI]	0.1033	0.1003	0.0215	8.7	23.2
Ec1	[DESOTF] [DCA]	0.1012	0.101	0.0245	9.13	25
Ec2	[DESOPF] [DCA]	0.1015	0.1007	0.0234	8.31	30.4
Ec3	[DESOTF] [TFSI]	0.1008	0.1006	0.0227	7.92	22.8
Ec4	[DESOPF] [TFSI]	0.1015	0.1002	0.0227	7.44	21.7

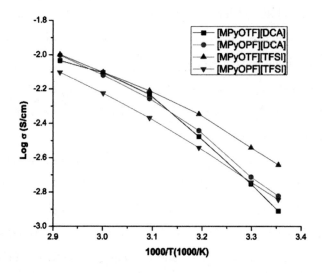

Figure 5.2. Ionic Conductivity versus Temperature of neat MPy-based FRTILs

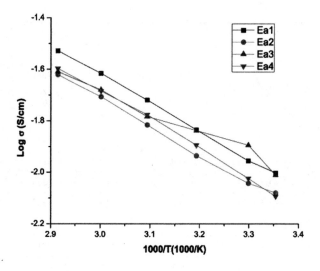

Figure 5.3. Ionic Conductivity versus Temperature of MPy-based electrolytes

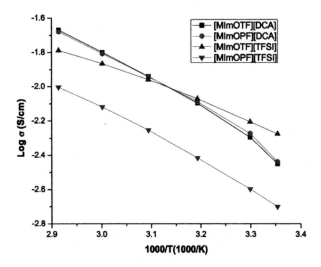

Figure 5.4. Ionic Conductivity versus Temperature of neat MIm-based FRTILs

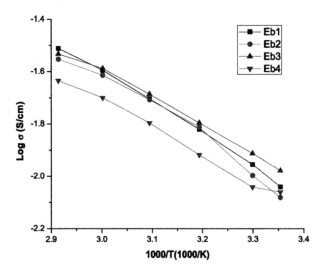

Figure 5.5. Ionic Conductivity versus Temperature of MIm-based electrolytes

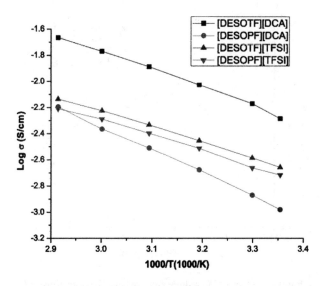

Figure 5.6. Ionic Conductivity versus Temperature of neat DES-based FRTILs

Many parameters can affect their conductivity, such as viscosity, density, ion size, anionic charge delocalization, aggregations and ionic motions. So it is relatively hard to estimate the contribution of each parameter to the conductivity of a liquid. However, proportionality between the conductivity and counter of the viscosity has been detected for numerous liquids in a varied temperature range. It is apparent that ionic conductivity increases with increasing temperature for all the investigated neat FRTILs and FRTILs based electrolytes. The decrease in viscosity [111] and faster migration of carrier ions and easier motion in the electrolyte system at higher temperatures are possibly responsible for this phenomenon [112]. As can be seen from Tables 5.2 & 5.4, such a trend appears almost to be varied [123].

Conductivity of the series of these neat FRTILs is on the order of 10^{-3} mS/cm at ambient temperatures. The ionic conductivity of the neat [MIm-

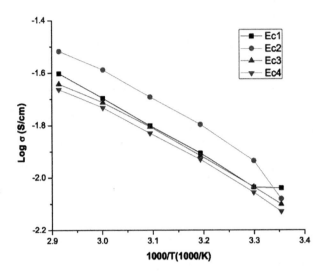

Figure 5.7. Ionic Conductivity versus Temperature of DES-based electrolytes

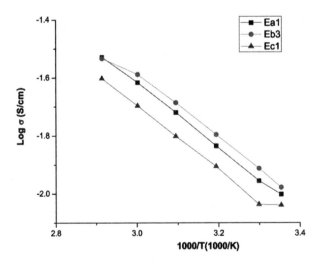

Figure 5.8. Ionic Conductivity versus Temperature of best three FRTILs based electrolytes

OTF] [TFSI] increases from 5.3 mS/cm at 25 ℃ and reaches 16.29 mS/cm at 75 ℃. When Eb2 is combined with 0.5 M LiTFSI in [MImOTF] [TFSI] / [MImEt] [DCA], as expected, a significant increase in the ionic conductivity is observed even at low temperatures. The ionic conductivity of Eb3 increases from 10.51 mS/cm at 25 ℃ to 29.3 mS/cm at 75 ℃. The effect of [MImEt] [DCA] on the improvement of the bulk conductivity lies in its ability to increase the number of dissociated ions LiTFSI in the solvent mixture by increasing its dielectric constant to values presumably higher than that of [MImOTF] [TFSI]. The addition of [MImEt] [DCA] is also able to enhance the mobility of those dissociated ions by decreasing the viscosity of the solvent mixture, and hence increase the bulk conductivity of the solvent mixture electrolytes, compared with the single FRTILs electrolytes [113].

Meanwhile, in the electrolyte system FRTILs serve not only as non-flammable and non-volatile plasticizers but also as ionic carriers, because FRTILs with an EO unit are good solvents for lithium salts (LiTFSI) forming homogeneous liquid solutions. So comparing neat FRTILs (Figures 5.2, 5.4, 5.6), the series of FRTILs based electrolytes (Figure 5.3, 5.5, 5.7) showed much higher conductivity.

Comparing Figures 5.3, 5.5, 5.7, the ionic conductivityies of the FRTILs terminated with-CF_3 group are totally higher than the ones of those terminated with-CF_2CF_2H group, for example Ea1 > Ea2, Eb1 > Eb2, Ec1 > Ec2. This interesting trend may be caused by the intermolecular hydrogen bonding between the-CF_2CF_2H group and the EO unit, which accounts for the observed higher viscosity and reduces the electrolyte system ordering. These results strongly imply that the ester group on the ionic liquid is playing a favorable role in transporting lithium ions through interaction with a lithium salt [111].

In addition, Figure 5.8 shows the best electrolytes of each series of FR-TILs. Eb3, which is imidazolium-based FRTILs electrolyte system, exhibits the highest ionic conductivity over the whole temperature range studied. The series of pyrrolidinium-based FRTILs electrolytes show a competitive ionic conductivity. And sulfonium cation-based FRTILs electrolytes show even lower ionic conductivity.

5.3.5 Cyclic Voltammetry

Figure 5.9 shows the cyclic voltammetry of best four FRTILs based electrolytes. As is shown, all the three electrolytes exhibit pretty similar high voltage stability. Ea1 shows the electrochemical stability in a potential range of $-2 \sim 4.0$ V; Eb3 exhibits in a range of $-1.6 \sim 3.8$ V and Ec1 performs in a range of $-1.5 \sim 4$ V.

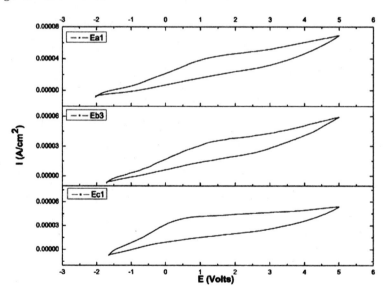

Figure 5.9. CV curves for the best three FRTILs based electrolytes

Ea1 has the widest electrochemicalwindow when the anodic potential is swept to the very positive region, an oxidation current arises at +4 V, suggesting the oxidative decomposition of the ILs system. This onset potential (+4 V) can be regarded as a "cathodic limit" of E8b, which defines the highest potential for the FRTILs-based electrolyte to be stable at. Obviously, this potential value of the cathodic limit is far positive than the highest charging potential (~3 V) of the sulfur cathode and therefore it is expected that not only can the Ea1 electrolyte be used for rechargeable sulfur cathodes, but also the other FRTILs based electrolyte systems. When the potential is scanned to the negative potential region, the CV curve shows a featureless background line until -2 V, indicating the chemical and electrochemical stability of the ILs system against lithium. With the presence of lithium ions in the electrolyte, the cathodic peak appeared at 1 V shows the electrochemical deposition of Li ions and the anodic peak appeared at -1.9 V indicates successive anodic oxidation of metallic lithium [116]. The appearance of a pair of oxidation-reduction peaks of lithium suggests that lithium can be very well charged and discharged in the Ea1 electrolyte system without any electrochemical hindrance. Except for these features, there are no any other current bands observed in the CV curve, showing a noble electrochemical stability in a potential range of -2 ~ 4 V and a sufficient potential window of the 0.5 M LiTFSI in [MPyOTF] [DCA] / [MImEt] [DCA] (Ea1) for Li/S cells [117]. The stability of the FRTILs based electrolytes to achieve high voltages against lithium is confirmed, which ensures their capability for further use in high potential electrolyte materials.

5.4 Summary

In summary, three series of FRTILs based on MPy, BIm and DES have been successfully synthesized and characterized. All prepared FRTILs exhibit good thermal stabilities up to 200 ℃ and non-flammability. Moreover, the ternary electrolytes of 0.5 M LiTFSI in FRTILs combined with [MImEt] [DCA] at 1 : 1 weight ratio have been prepared, and their properties have been investigated. The FRTILs based electrolytes show wide electrochemical stability and excellent ion conductivity (much more than the minimum requirement, which is >6mS/cm). This research establishes that the imidazolium-based FRTILs are promising candidates to replace volatile solvents due to their chemical and thermal stability, non-volatility, high ionic conductivity, large electrochemical window and good solvent behavior. Such impressive results, especially excellent battery performance at low-medium temperatures, also make the FRTILs based electrolytes prepared in this work promising electrolytes for very safe lithium rechargeable batteries.

Chapter 6   Conclusion

In this thesis, two types of rechargeable lithium batteries, lithium-ion and lithium-sulfur, are discussed. Two different approaches have been presented to develop superior electrolytes for rechargeable lithium batteries. One approach is based on the conventional PEO-based SPE system. The key feature of this approach is the preparation of NPLS and low lattice energy fluorinated dilithium salts. The ionic conductivities and electrochemical stability of these PEO-based SPEs are significantly improved.

For lithium-sulfur (Li-S) batteries, the polysulfide shuttle, caused by the dissolution of cathode polysulfide intermediates into the electrolyte, has delivered a mortal blow to nearly every attempt at obtaining a viable Li-S battery. So, another approach involves the strategic design and synthesis of a series of RTILs to prevent PSS: i) TFSI-based ILs are non-flammable, high electrochemical stability and thermostable. ii) DCA-based ILs have much lower viscosity than TFSI-based ILs and display much better ionic conductivity—by using DCA as the counter anion for the ILs. iii) Asymmetric fluorinated-based RTILs display excellent ionic conductivity, and adequate electrochemical & thermal stability.

For the characterization, we used FTIR, ^{1}H-NMR, and ^{13}C-NMR to confirm the structures of all compounds, and used viscosity and TGA to study the physical/thermal properties. We have conducted ionic conductivity and CV measurement to study the electrochemical properties of these materials. Finally, as part of the future direction of our research group, we still need to perform a collaborative work with other research groups using the best samples with the intention of obtaining more results, viz., test cell performance, charge/discharge, cycling, etc.

APPENDIX
FTIR AND NMR SPECTRA

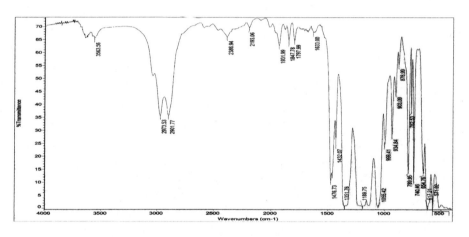

Figure A. 1. FTIR Spectrum of [DiMPy1O] [TFSI]

<<< APPENDIX FTIR AND NMR SPECTRA

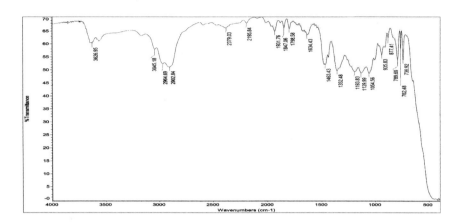

Figure A. 2. FTIR Spectrum of ［DiMPy2O］［TFSI］

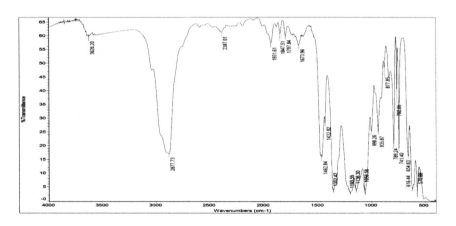

Figure A. 3. FTIR Spectrum of ［DiMPy3O］［TFSI］

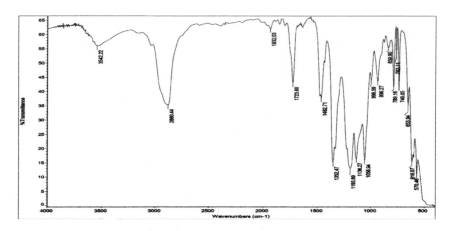

Figure A. 4. FTIR Spectrum of [DiMPy4O] [TFSI]

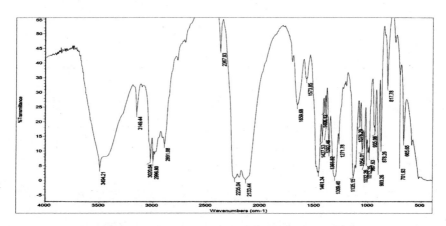

Figure A. 5. FTIR Spectrum of [DiMPy1O] [DCA]

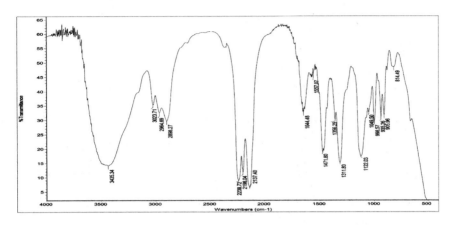

Figure A. 6. FTIR Spectrum of [DiMPy2O] [DCA]

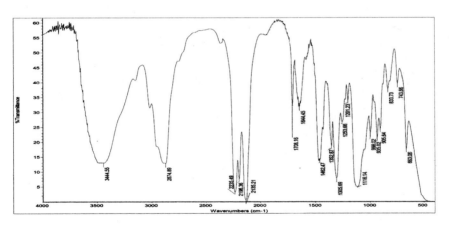

Figure A. 7. FTIR Spectrum of [DiMPy3O] [DCA]

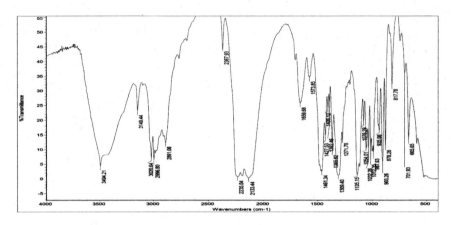

Figure A. 8. FTIR Spectrum of [DiMPy4O] [DCA]

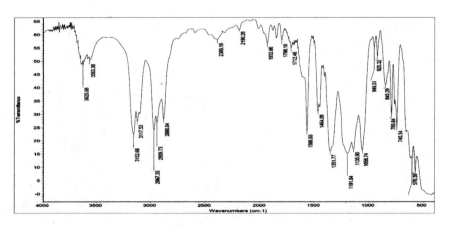

Figure A. 9. FTIR Spectrum of [DiBIm1O] [TFSI]

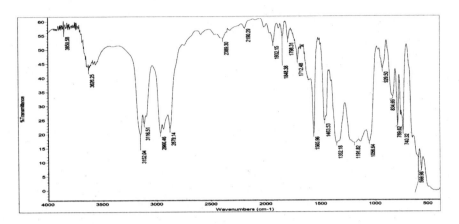

Figure A. 10. FTIR Spectrum of [DiBIm2O] [TFSI]

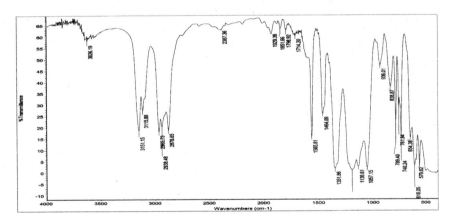

Figure A. 11. FTIR Spectrum of [DiBIm3O] [TFSI]

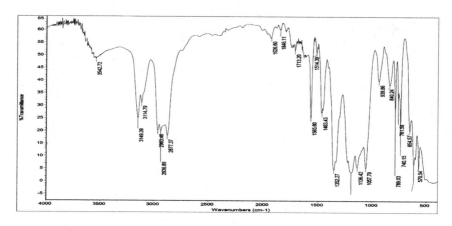

Figure A. 12. FTIR Spectrum of [DiBIm4O] [TFSI]

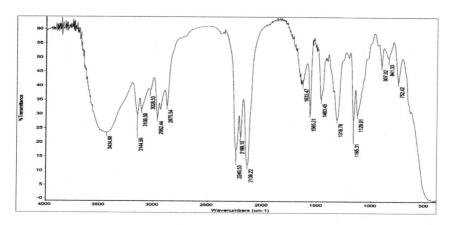

Figure A. 13. FTIR Spectrum of [DiBIm1O] [DCA]

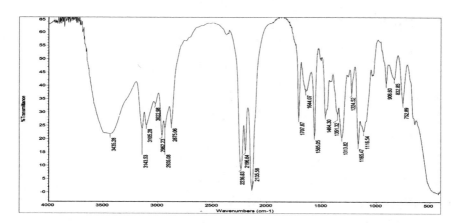

Figure A. 14. FTIR Spectrum of [DiBIm2O][DCA]

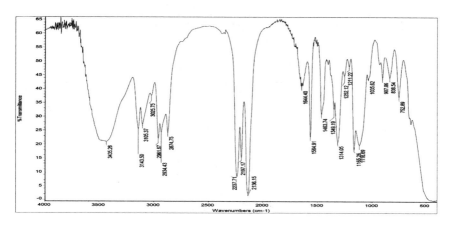

Figure A. 15. FTIR Spectrum of [DiBIm3O][DCA]

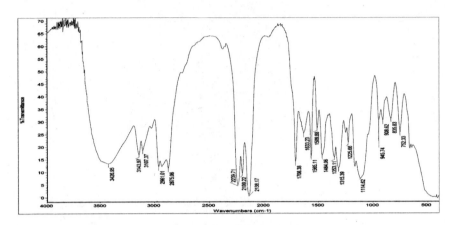

Figure A. 16. FTIR Spectrum of [DiBIm4O] [DCA]

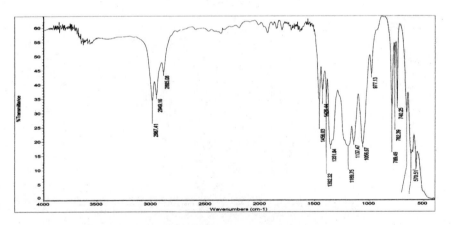

Figure A. 17. FTIR Spectrum of [DiDES1O] [TFSI]

<<< APPENDIX FTIR AND NMR SPECTRA

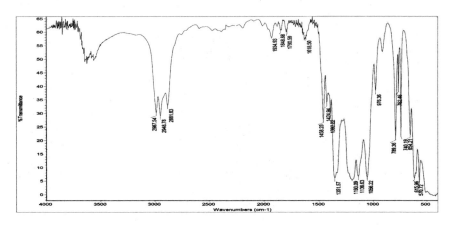

Figure A. 18. FTIR Spectrum of [DiDES2O] [TFSI]

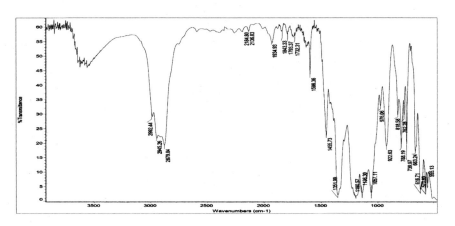

Figure A. 19. FTIR Spectrum of [DiDES3O] [TFSI]

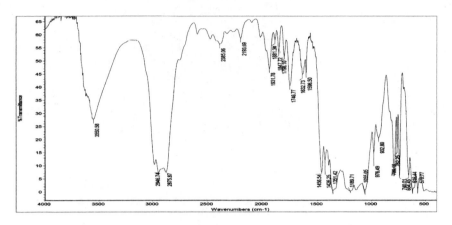

Figure A. 20. FTIR Spectrum of ［DiDES4O］［TFSI］

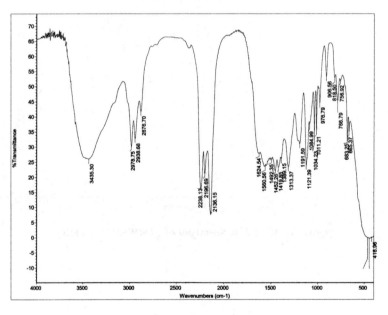

Figure A. 21. FTIR Spectrum of ［DiDES1O］［DCA］

<<< APPENDIX FTIR AND NMR SPECTRA

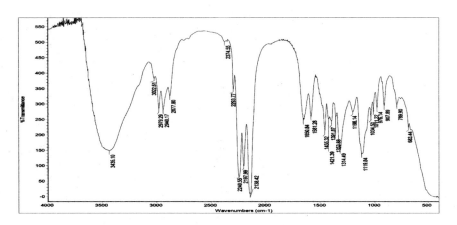

Figure A. 22. FTIR Spectrum of [DiDES2O] [DCA]

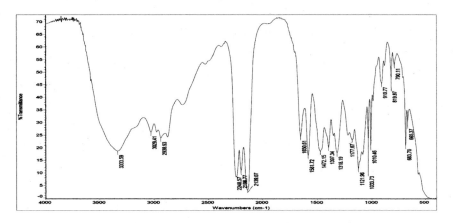

Figure A. 23. FTIR Spectrum of [DiDES3O] [DCA]

159

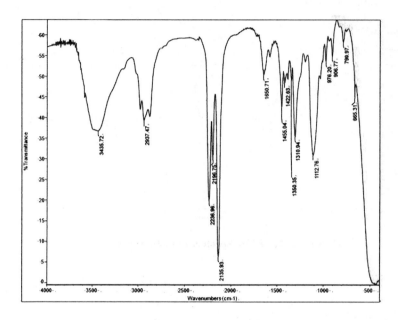

Figure A. 24. FTIR Spectrum of [DiDES4O] [DCA]

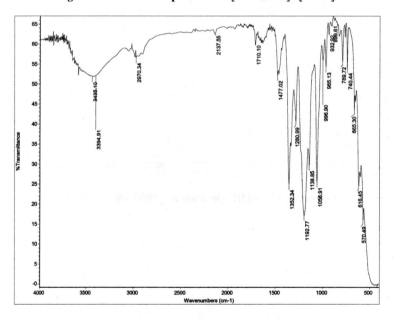

Figure A. 25. FTIR Spectrum of [MPyOTF] [TFSI]

<<< APPENDIX FTIR AND NMR SPECTRA

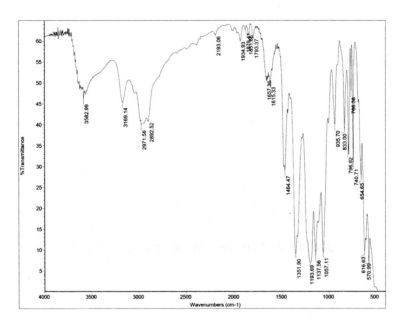

Figure A. 26. FTIR Spectrum of [MPyOPF] [TFSI]

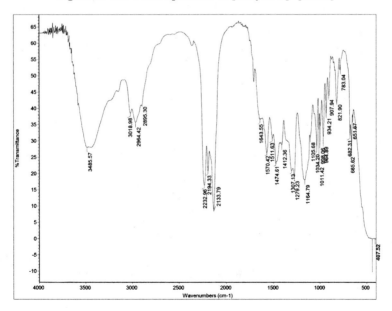

Figure A. 27. FTIR Spectrum of [MPyOTF] [DCA]

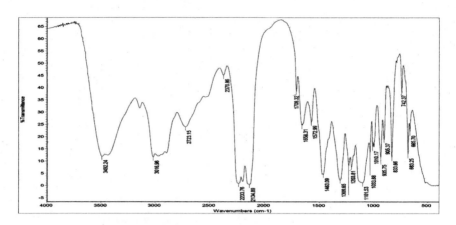

Figure A. 28. FTIR Spectrum of [MPyOPF] [DCA]

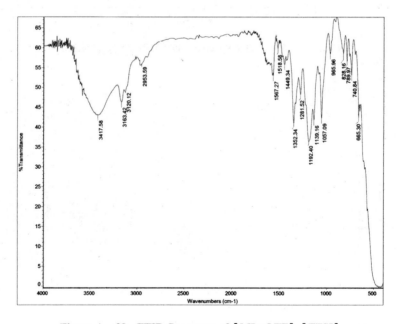

Figure A. 29. FTIR Spectrum of [MImOTF] [TFSI]

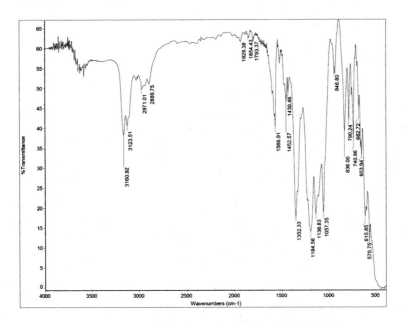

Figure A. 30. FTIR Spectrum of [MImOPF] [TFSI]

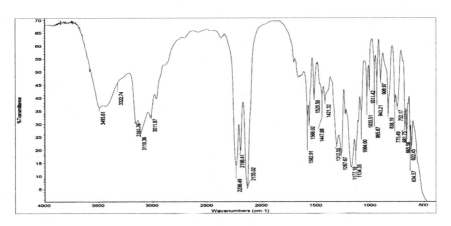

Figure A. 31. FTIR Spectrum of [MPyOTF] [DCA]

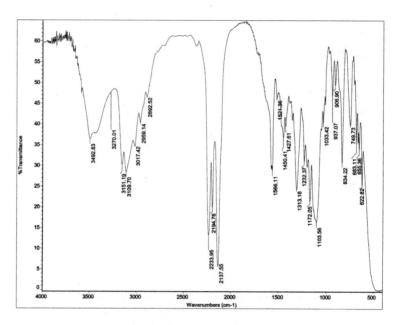

Figure A. 32. FTIR Spectrum of [MPyOPF] [DCA]

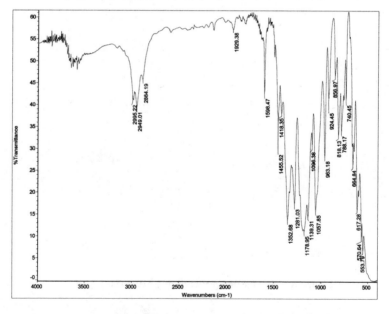

Figure A. 33. FTIR Spectrum of [DESOTF] [TFSI]

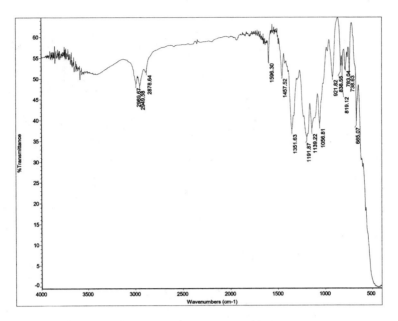

Figure A. 34. FTIR Spectrum of [DESOPF] [TFSI]

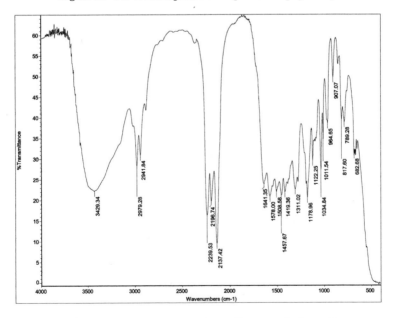

Figure A. 35. FTIR Spectrum of [DESOTF] [DCA]

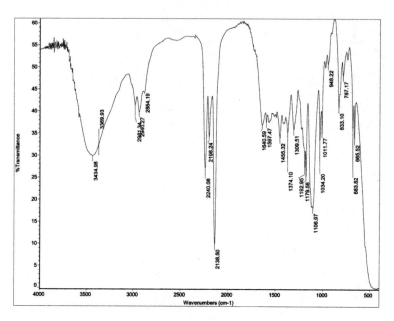

Figure A. 36. FTIR Spectrum of [DESOPF] [DCA]

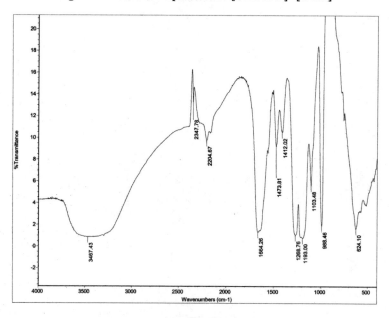

Figure A. 37. FTIR Spectrum of LS – 1

<<< APPENDIX FTIR AND NMR SPECTRA

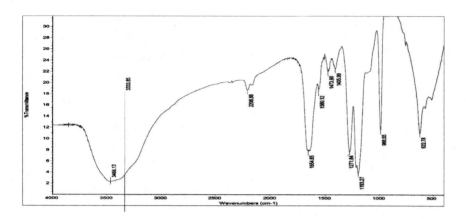

Figure A. 38. FTIR Spectrum of LS – 2

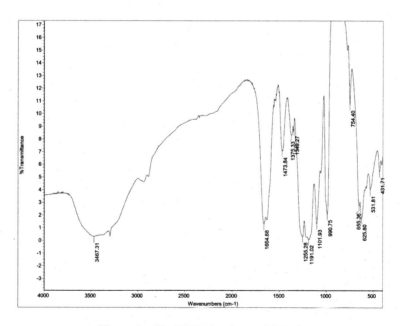

Figure A. 39. FTIR Spectrum of LS – 3

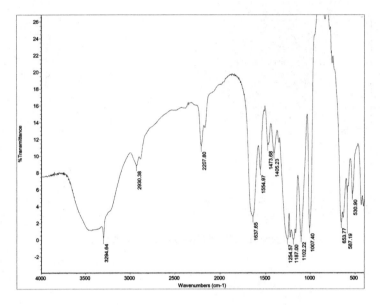

Figure A. 40. FTIR Spectrum of LS – 4

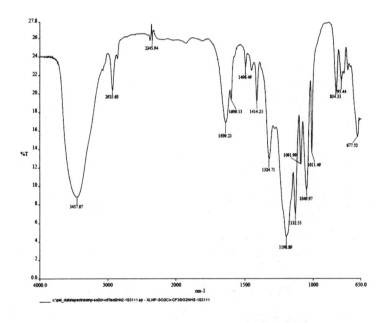

Figure A. 41. FTIR Spectrum of PSTFSILi

<<< APPENDIX FTIR AND NMR SPECTRA

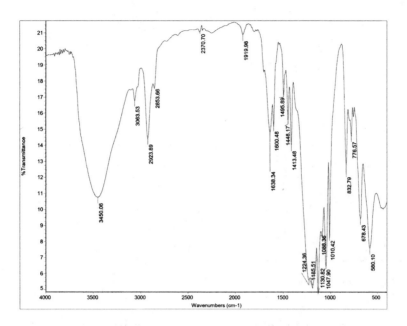

Figure A. 42. FTIR Spectrum of PSPhSILi

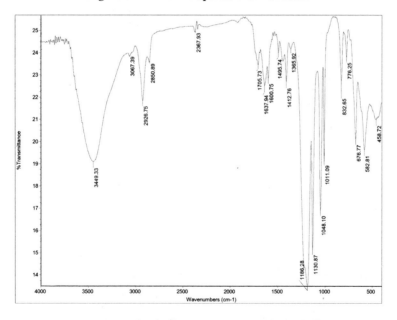

Figure A. 43. FTIR Spectrum of PSDTTOLi

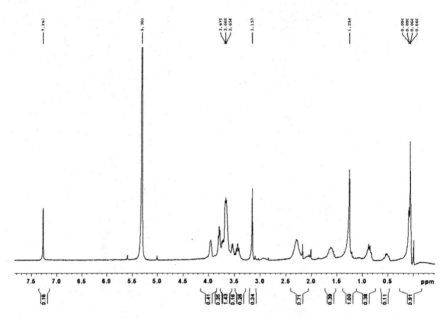

Figure B. 1. HNMR Spectrum of [DiMPy1O] [TFSI]

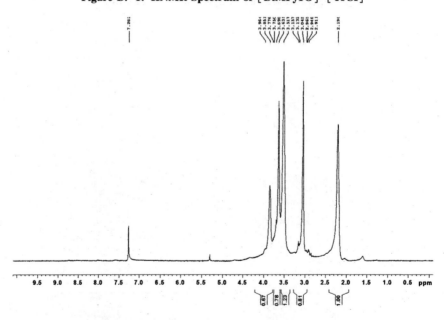

Figure B. 2. HNMR Spectrum of [DiMPy2O] [TFSI]

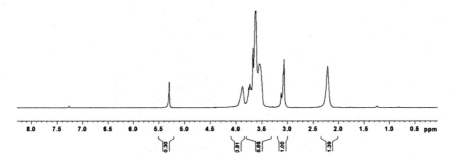

Figure B. 3. HNMR Spectrum of [DiMPy3O] [TFSI]

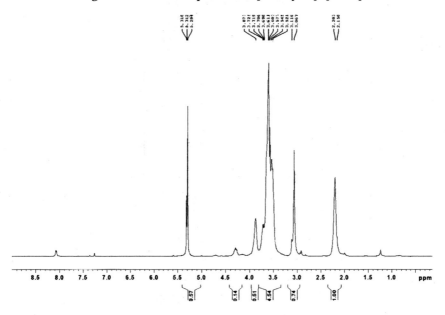

Figure B. 4. HNMR Spectrum of [DiMPy4O] [TFSI]

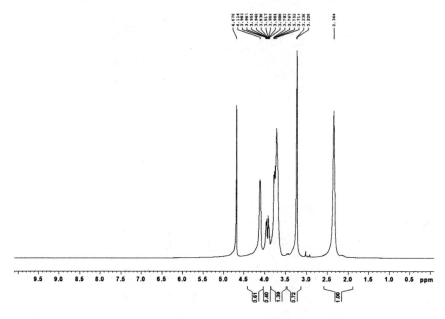

Figure B. 5. HNMR Spectrum of [DiMPy1O] [DCA]

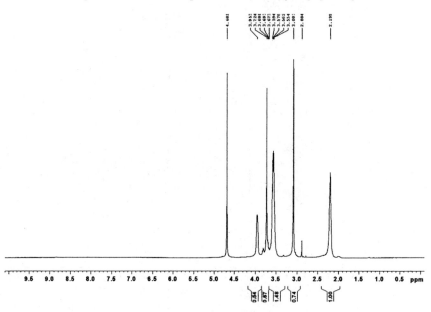

Figure B. 6. HNMR Spectrum of [DiMPy2O] [DCA]

<<< APPENDIX FTIR AND NMR SPECTRA

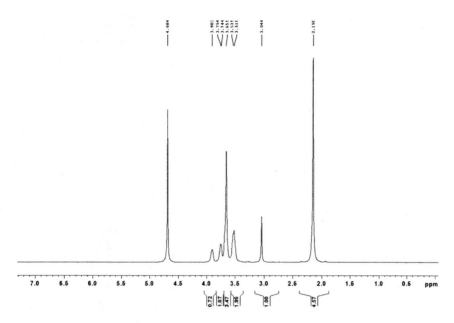

Figure B. 7. HNMR Spectrum of [DiMPy3O][DCA]

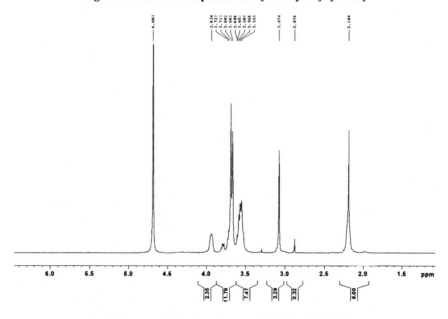

Figure B. 8. HNMR Spectrum of [DiMPy4O][DCA]

173

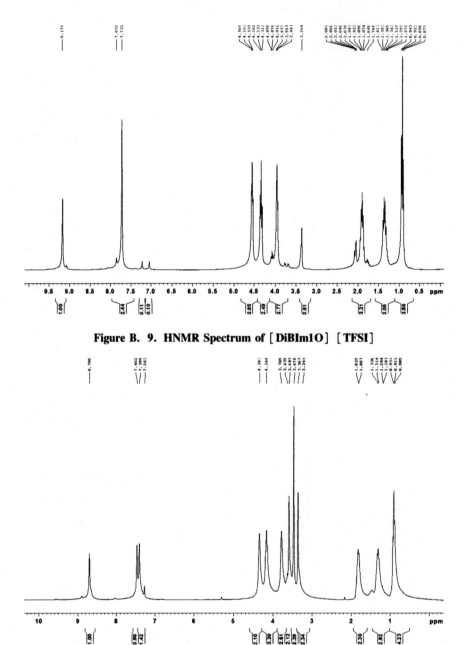

Figure B. 9. HNMR Spectrum of [DiBIm1O] [TFSI]

Figure B. 10. HNMR Spectrum of [DiBIm2O] [TFSI]

<<< APPENDIX FTIR AND NMR SPECTRA

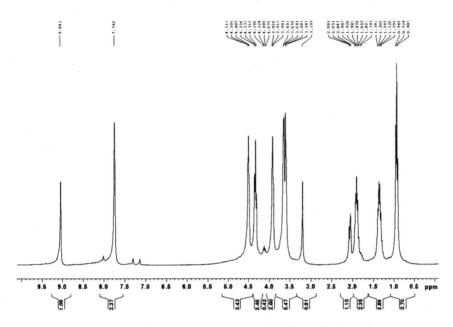

Figure B. 11. HNMR Spectrum of [DiBIm3O] [TFSI]

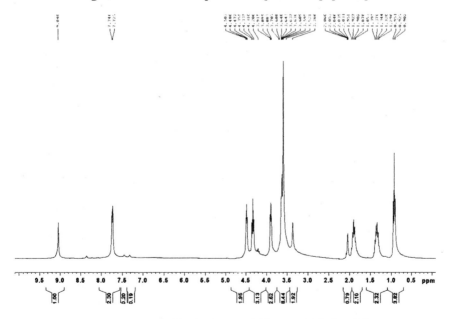

Figure B. 12. HNMR Spectrum of [DiBIm4O] [TFSI]

175

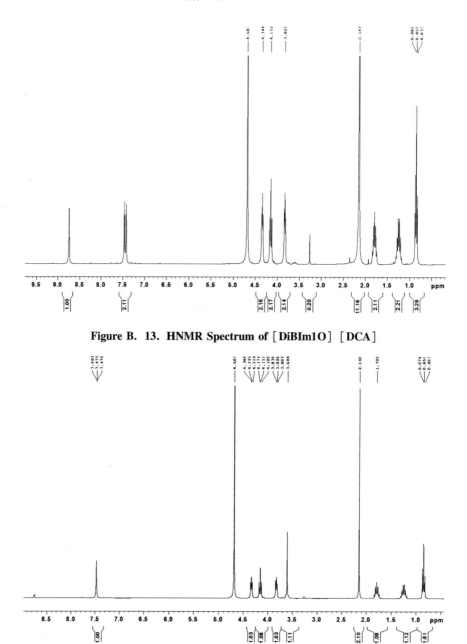

Figure B. 13. HNMR Spectrum of [DiBIm1O] [DCA]

Figure B. 14. HNMR Spectrum of [DiBIm2O] [DCA]

<<< APPENDIX FTIR AND NMR SPECTRA

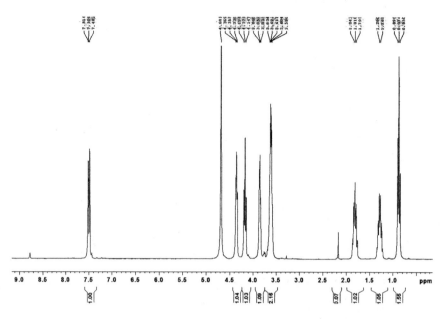

Figure B. 15. HNMR Spectrum of [DiBIm3O][DCA]

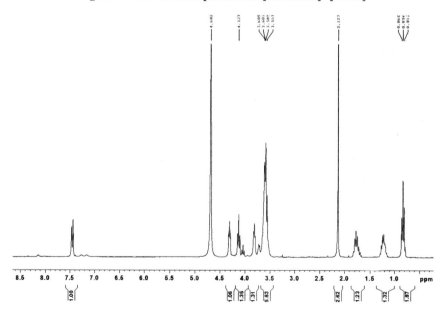

Figure B. 16. HNMR Spectrum of [DiBIm4O][DCA]

177

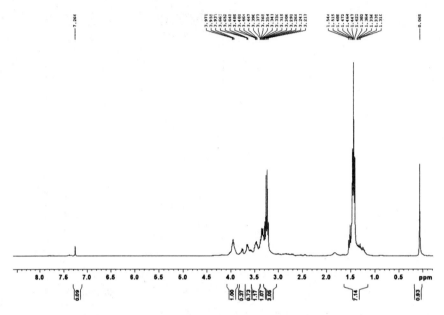

Figure B. 17. HNMR Spectrum of [DiDES1O] [TFSI]

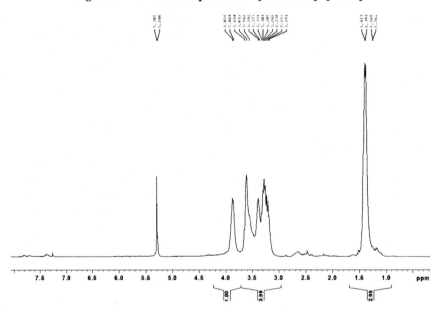

Figure B. 18. HNMR Spectrum of [DiDES2O] [TFSI]

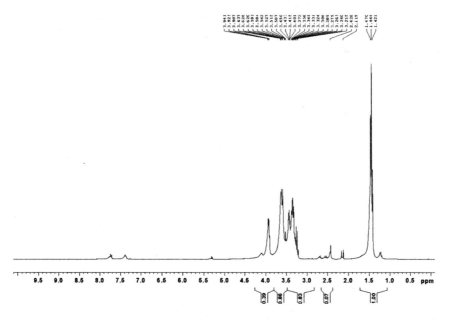

Figure B. 19. HNMR Spectrum of [DiDES3O] [TFSI]

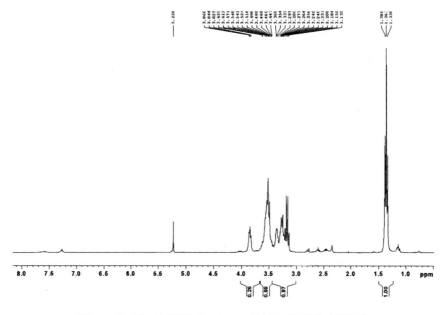

Figure B. 20. HNMR Spectrum of [DiDES4O] [TFSI]

179

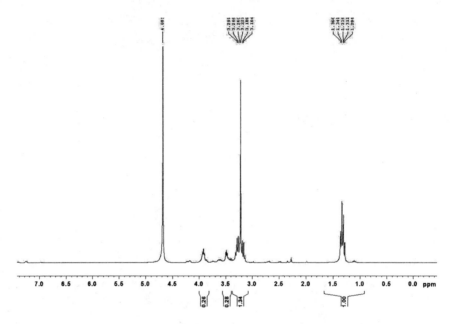

Figure B. 21. HNMR Spectrum of [DiDES1O] [DCA]

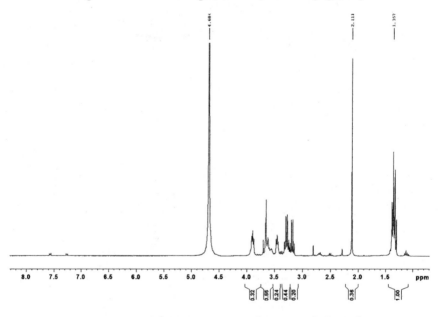

Figure B. 22. HNMR Spectrum of [DiDES2O] [DCA]

<<< APPENDIX FTIR AND NMR SPECTRA

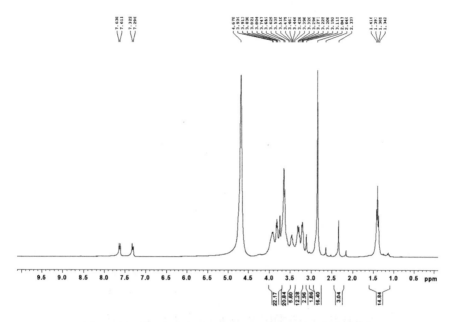

Figure B. 23. HNMR Spectrum of [DiDES3O][DCA]

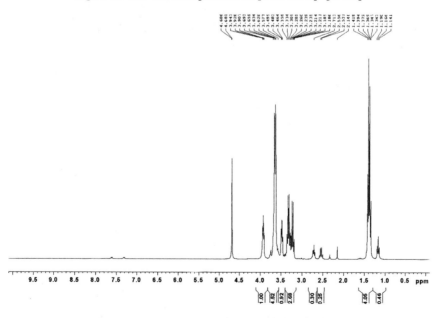

Figure B. 24. HNMR Spectrum of [DiDES4O][DCA]

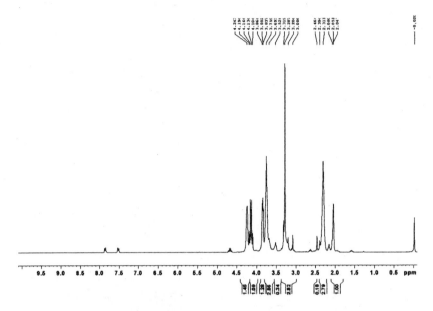

Figure B. 25. HNMR Spectrum of [MPyOTF] [TFSI]

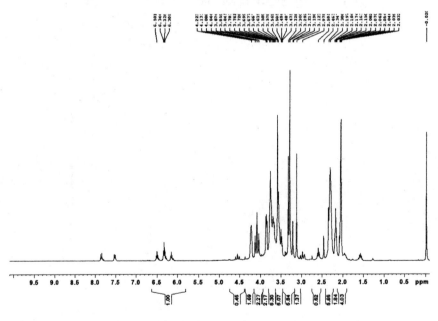

Figure B. 26. HNMR Spectrum of [MPyOPF] [TFSI]

<<< APPENDIX　FTIR AND NMR SPECTRA

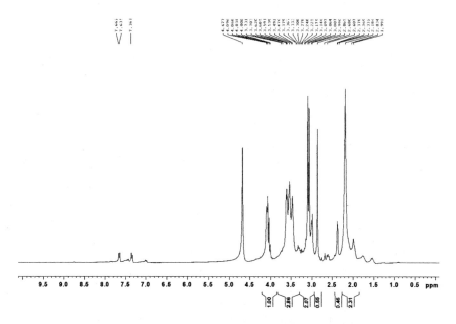

Figure B. 27. HNMR Spectrum of [MPyOTF] [DCA]

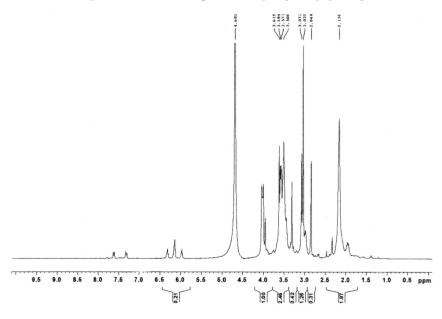

Figure B. 28. HNMR Spectrum of [MPyOPF] [DCA]

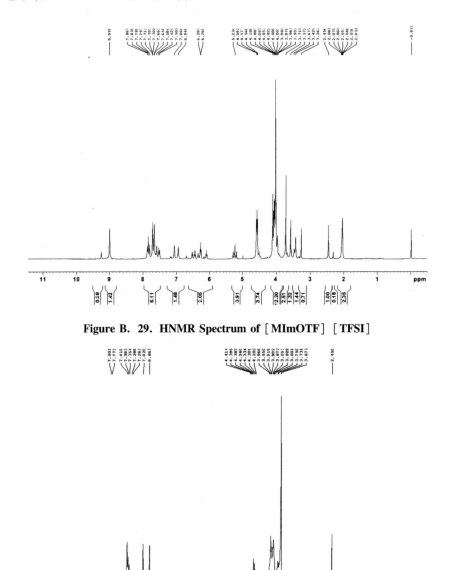

Figure B. 29. HNMR Spectrum of [MImOTF] [TFSI]

Figure B. 30. HNMR Spectrum of [MImOPF] [TFSI]

<<< APPENDIX FTIR AND NMR SPECTRA

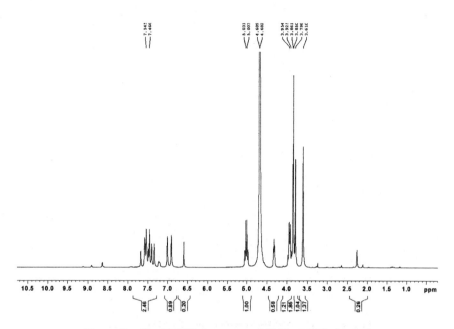

Figure B. 31. HNMR Spectrum of [MImOTF] [DCA]

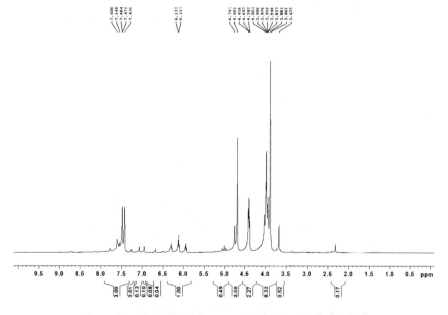

Figure B. 32. HNMR Spectrum of [MImOPF] [DCA]

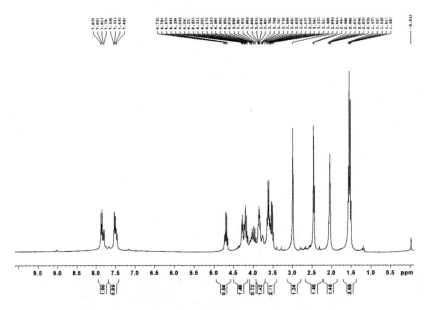

Figure B. 33. HNMR Spectrum of [DESOTF] [TFSI]

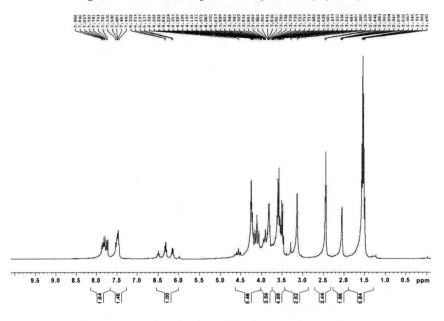

Figure B. 34. HNMR Spectrum of [DESOPF] [TFSI]

<<< APPENDIX FTIR AND NMR SPECTRA

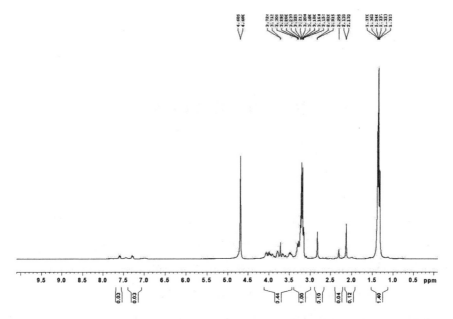

Figure B. 35. HNMR Spectrum of [DESOTF] [DCA]

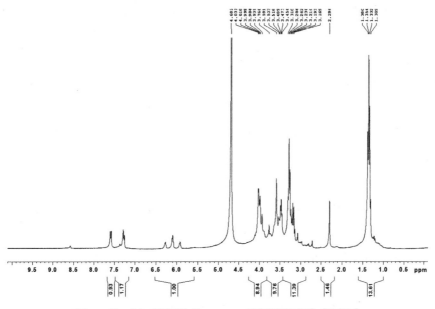

Figure B. 36. HNMR Spectrum of [DESOPF] [DCA]

BIBLIOGRAPHY

[1] J. M. Tarascon, A. S. Gozdz, C. Schmutz, F. Shokoohi, and P. C. Warren, "Performance of Bellcore's plastic rechargeable Li-ion batteries," *Solid State Ion.*, vol. 86–88, pp. 49–54, Jul. 1996.

[2] A. R. Polu and H. -W. Rhee, "Nanocomposite solid polymer electrolytes based on poly (ethylene oxide) /POSS-PEG (n = 13.3) hybrid nanoparticles for lithium ion batteries," *J. Ind. Eng. Chem.*, vol. 31, pp. 323–329, Nov. 2015.

[3] Z. Wang, Y. Fu, Z. Zhang, S. Yuan, K. Amine, V. Battaglia, and G. Liu, "Application of Stabilized Lithium Metal Powder (SLMP®) in graphite anode-A high efficient prelithiation method for lithium-ion batteries," *J. Power Sources*, vol. 260, pp. 57–61, Aug. 2014.

[4] P. Joge, D. K. Kanchan, P. Sharma, M. Jayswal, and D. K. Avasthi, "Effect of swift heavy O7 + ion radiations on conductivity of lithium based polymer blend electrolyte," *Radiat. Phys. Chem.*, vol. 100, pp. 74–79, Jul. 2014.

[5] A. Chakrabarti, R. Filler, and B. K. Mandal, "Borate ester plasticizer for PEO-based solid polymer electrolytes," *J. Solid State Electrochem.*, vol. 12, no. 3, pp. 269–272, Mar. 2008.

[6] P. Zare, M. Mahrova, E. Tojo, A. Stojanovic, and W. H. Binder, "Ethylene glycol-based ionic liquids via azide/alkyne click chemistry," *J. Polym. Sci. Part Polym. Chem.*, vol. 51, no. 1, pp. 190 – 202, Jan. 2013.

[7] X. Fang and H. Peng, "A Revolution in Electrodes: Recent Progress in Rechargeable Lithium-Sulfur Batteries," *Small*, vol. 11, no. 13, pp. 1488 – 1511, Apr. 2015.

[8] J. Scheers, S. Fantini, and P. Johansson, "A review of electrolytes for lithium-sulphur batteries," *J. Power Sources*, vol. 255, pp. 204 – 218, Jun. 2014.

[9] R. M. McMeeking and R. Purkayastha, "The Role of Solid Mechanics in Electrochemical Energy Systems such as Lithium-ion Batteries," *Procedia IUTAM*, vol. 10, pp. 294 – 306, 2014.

[10] K. M. Kim, B. Z. Poliquit, Y. -G. Lee, J. Won, J. M. Ko, and W. I. Cho, "Enhanced separator properties by thermal curing of poly (ethylene glycol) diacrylate-based gel polymer electrolytes for lithium-ion batteries," *Electrochimica Acta*, vol. 120, pp. 159 – 166, Feb. 2014.

[11] Q. Cheng, Z. Cui, J. Li, S. Qin, F. Yan, and J. Li, "Preparation and performance of polymer electrolyte based on poly (vinylidene fluoride) /polysulfone blend membrane via thermally induced phase separation process for lithium ion battery," *J. Power Sources*, vol. 266, pp. 401 – 413, Nov. 2014.

[12] J. Zhou, "Ionic conductivity of composite electrolytes based on oligo (ethylene oxide) and fumed oxides," *Solid State Ion.*, vol. 166, no. 3 – 4, pp. 275 – 293, Jan. 2004.

[13] Q. Lu, J. Fang, J. Yang, G. Yan, S. Liu, and J. Wang,

"A novel solid composite polymer electrolyte based on poly (ethylene oxide) segmented polysulfone copolymers for rechargeable lithium batteries," *J. Membr. Sci.*, vol. 425–426, pp. 105–112, Jan. 2013.

[14] Q. Wang, H. Fan, L.-Z. Fan, and Q. Shi, "Preparation and performance of a non-ionic plastic crystal electrolyte with the addition of polymer for lithium ion batteries," *Electrochimica Acta*, vol. 114, pp. 720–725, Dec. 2013.

[15] A. M. Haregewoin, E. G. Leggesse, J.-C. Jiang, F.-M. Wang, B.-J. Hwang, and S. D. Lin, "Comparative Study on the Solid Electrolyte Interface Formation by the Reduction of Alkyl Carbonates in Lithium ion Battery," *Electrochimica Acta*, vol. 136, pp. 274–285, Aug. 2014.

[16] J. Scheers, S. Fantini, and P. Johansson, "A review of electrolytes for lithium-sulphur batteries," *J. Power Sources*, vol. 255, pp. 204–218, Jun. 2014.

[17] R. C. Agrawal and G. P. Pandey, "Solid polymer electrolytes: materials designing and all-solid-state battery applications: an overview," *J. Phys. Appl. Phys.*, vol. 41, no. 22, p. 223001, Nov. 2008.

[18] M. Anderman, "Lithium-polymer batteries for electrical vehicles: A realistic view," *Solid State Ion.*, vol. 69, no. 3–4, pp. 336–342, Aug. 1994.

[19] R. Schmitz, R. Schmitz, R. Müller, O. Kazakova, N. Kalinovich, G.-V. Röschenthaler, M. Winter, S. Passerini, and A. Lex-Balducci, "Methyl tetrafluoro-2-(methoxy) propionate as co-solvent for propylene carbonate-based electrolytes for lithium-ion batteries," *J. Power Sources*, vol. 205, pp. 408–413, May 2012.

[20] R. Chen, L. Zhu, F. Wu, L. Li, R. Zhang, and S. Chen, "Investigation of a novel ternary electrolyte based on dimethyl sulfite and lithium difluoromono (oxalato) borate for lithium ion batteries," *J. Power Sources*, vol. 245, pp. 730–738, Jan. 2014.

[21] S. D. Tillmann, P. Isken, and A. Lex-Balducci, "Gel polymer electrolyte for lithium-ion batteries comprising cyclic carbonate moieties," *J. Power Sources*, vol. 271, pp. 239–244, Dec. 2014.

[22] H. Zhang, C. Liu, L. Zheng, F. Xu, W. Feng, H. Li, X. Huang, M. Armand, J. Nie, and Z. Zhou, "Lithium bis (fluorosulfonyl) imide/poly (ethylene oxide) polymer electrolyte," *Electrochimica Acta*, vol. 133, pp. 529–538, Jul. 2014.

[23] Y. Xing, Y. Wu, H. Wang, G. Yang, W. Li, L. Xu, and X. Jiang, "Preparation of hybrid polymer based on polyurethane lithium salt and polyvinylidene fluoride as electrolyte for lithium-ion batteries," *Electrochimica Acta*, vol. 136, pp. 513–520, Aug. 2014.

[24] P. V. Wright, "Electrical conductivity in ionic complexes of poly (ethylene oxide)," *Br. Polym. J.*, vol. 7, no. 5, pp. 319–327, Sep. 1975.

[25] D. E. Fenton, J. M. Parker, and P. V. Wright, "Complexes of alkali metal ions with poly (ethylene oxide)," *Polymer*, vol. 14, no. 11, p. 589, Nov. 1973.

[26] X. Wang, C. Gong, D. He, Z. Xue, C. Chen, Y. Liao, and X. Xie, "Gelled microporous polymer electrolyte with low liquid leakage for lithium-ion batteries," *J. Membr. Sci.*, vol. 454, pp. 298–304, Mar. 2014.

[27] M. Marzantowicz, J. R. Dygas, F. Krok, A. Tomaszewska, Z. Florjańczyk, E. Zygadło-Monikowska, and G. Lapienis, "Star-

branched poly (ethylene oxide) LiN (CF$_3$SO$_2$)$_2$: A promising polymer electrolyte," *J. Power Sources*, vol. 194, no. 1, pp. 51 –57, Oct. 2009.

[28] A. S. Aricò, P. Bruce, B. Scrosati, J. -M. Tarascon, and W. van Schalkwijk, "Nanostructured materials for advanced energy conversion and storage devices," *Nat. Mater.*, vol. 4, no. 5, pp. 366 –377, May 2005.

[29] P. A. R. Jayathilaka, M. A. K. Dissanayake, I. Albinsson, and B. -E. Mellander, "Effect of nano-porous Al$_2$O$_3$ on thermal, dielectric and transport properties of the (PEO) 9LiTFSI polymer electrolyte system," *Electrochimica Acta*, vol. 47, no. 20, pp. 3257 –3268, Aug. 2002.

[30] B. Liang, S. Tang, Q. Jiang, C. Chen, X. Chen, S. Li, and X. Yan, "Preparation and characterization of PEO-PMMA polymer composite electrolytes doped with nano-Al2O3," *Electrochimica Acta*, vol. 169, pp. 334 –341, Jul. 2015.

[31] S. I. Moon, C. R. Lee, B. S. Jin, K. E. Min, and W. S. Kim, "Ionic conductivities of cross-linked polymer electrolytes prepared from oligo (ethylene glycol) dimethacrylates," *J. Power Sources*, vol. 87, no. 1 –2, pp. 223 –225, Apr. 2000.

[32] H. Cheng, C. Zhu, B. Huang, M. Lu, and Y. Yang, "Synthesis and electrochemical characterization of PEO-based polymer electrolytes with room temperature ionic liquids," *Electrochimica Acta*, vol. 52, no. 19, pp. 5789 –5794, May 2007.

[33] A. Asghar, Y. Abdul Samad, B. Singh Lalia, and R. Hashaikeh, "PEG based quasi-solid polymer electrolyte: Mechanically supported by networked cellulose," *J. Membr. Sci.*, vol. 421 –422, pp. 85 –90, Dec. 2012.

[34] C. M. Costa, J. L. Gomez Ribelles, S. Lanceros-Méndez, G. B. Appetecchi, and B. Scrosati, "Poly (vinylidene fluoride) -based, co-polymer separator electrolyte membranes for lithium-ion battery systems," *J. Power Sources*, vol. 245, pp. 779–786, Jan. 2014.

[35] M. Marzantowicz, J. R. Dygas, F. Krok, A. Łasińska, Z. Florjańczyk, E. Zygadło-Monikowska, and A. Affek, "Crystallization and melting of PEO: LiTFSI polymer electrolytes investigated simultaneously by impedance spectroscopy and polarizing microscopy," *Electrochimica Acta*, vol. 50, no. 19, pp. 3969–3977, Jun. 2005.

[36] A. Marrocchi, P. Adriaensens, E. Bartollini, B. Barkakaty, R. Carleer, J. Chen, D. K. Hensley, C. Petrucci, M. Tassi, and L. Vaccaro, "Novel cross-linked polystyrenes with large space network as tailor-made catalyst supports for sustainable media," *Eur. Polym. J.*, vol. 73, pp. 391–401, Dec. 2015.

[37] X. Ji, K. T. Lee, and L. F. Nazar, "A highly ordered nanostructured carbon-sulphur cathode for lithium-sulphur batteries," *Nat. Mater.*, vol. 8, no. 6, pp. 500–506, Jun. 2009.

[38] L. Chen and L. L. Shaw, "Recent advances in lithium-sulfur batteries," *J. Power Sources*, vol. 267, pp. 770–783, Dec. 2014.

[39] D.-W. Wang, Q. Zeng, G. Zhou, L. Yin, F. Li, H.-M. Cheng, I. R. Gentle, and G. Q. M. Lu, "Carbon-sulfur composites for Li-S batteries: status and prospects," *J. Mater. Chem. A*, vol. 1, no. 33, p. 9382, 2013.

[40] F. Wu, Q. Zhu, R. Chen, N. Chen, Y. Chen, and L. Li, "A Safe Electrolyte with Counterbalance between the Ionic Liquid and Tris (ethylene glycol) dimethyl ether for High Performance Lithium-Sulfur Batteries," *Electrochimica Acta*, vol. 184, pp. 356–363, Dec. 2015.

[41] M. -K. Song, E. J. Cairns, and Y. Zhang, "Lithium/sulfur batteries with high specific energy: old challenges and new opportunities," *Nanoscale*, vol. 5, no. 6, p. 2186, 2013.

[42] D. Brouillette, G. Perron, and J. E. Desnoyers, "Effect of viscosity and volume on the specific conductivity of lithium salts in solvent mixtures," *Electrochimica Acta*, vol. 44, no. 26, pp. 4721 – 4742, Sep. 1999.

[43] N. Azimi, Z. Xue, I. Bloom, M. L. Gordin, D. Wang, T. Daniel, C. Takoudis, and Z. Zhang, "Understanding the Effect of a Fluorinated Ether on the Performance of Lithium-Sulfur Batteries," *ACS Appl. Mater. Interfaces*, vol. 7, no. 17, pp. 9169 – 9177, May 2015.

[44] T. Achiha, T. Nakajima, Y. Ohzawa, M. Koh, A. Yamauchi, M. Kagawa, and H. Aoyama, "Thermal Stability and Electrochemical Properties of Fluorine Compounds as Nonflammable Solvents for Lithium-Ion Batteries," *J. Electrochem. Soc.*, vol. 157, no. 6, p. A707, 2010.

[45] C. Zu, N. Azimi, Z. Zhang, and A. Manthiram, "Insight into lithium-metal anodes in lithium-sulfur batteries with a fluorinated ether electrolyte," *J Mater Chem A*, vol. 3, no. 28, pp. 14864 – 14870, 2015.

[46] P. G. Bruce, S. A. Freunberger, L. J. Hardwick, and J. -M. Tarascon, "Li-O2 and Li-S batteries with high energy storage," *Nat. Mater.*, vol. 11, no. 1, pp. 19 – 29, Dec. 2011.

[47] N. Jayaprakash, J. Shen, S. S. Moganty, A. Corona, and L. A. Archer, "Porous Hollow Carbon@ Sulfur Composites for High-Power Lithium-Sulfur Batteries," *Angew. Chem.*, vol. 123, no. 26, pp. 6026 – 6030, Jun. 2011.

[48] J. Wang, S. Y. Chew, Z. W. Zhao, S. Ashraf, D. Wexler, J. Chen, S. H. Ng, S. L. Chou, and H. K. Liu, "Sulfur-mesoporous carbon composites in conjunction with a novel ionic liquid electrolyte for lithium rechargeable batteries," *Carbon*, vol. 46, no. 2, pp. 229 – 235, Feb. 2008.

[49] J. S. Wilkes and M. J. Zaworotko, "Air and water stable 1-ethyl-3-methylimidazolium based ionic liquids," *J. Chem. Soc. Chem. Commun.*, no. 13, p. 965, 1992.

[50] M. Mahrova, F. Pagano, V. Pejakovic, A. Valea, M. Kalin, A. Igartua, and E. Tojo, "Pyridinium based dicationic ionic liquids as base lubricants or lubricant additives," *Tribol. Int.*, vol. 82, pp. 245 – 254, Feb. 2015.

[51] Q. Q. Baltazar, J. Chandawalla, K. Sawyer, and J. L. Anderson, "Interfacial and micellar properties of imidazolium-based monocationic and dicationic ionic liquids," *Colloids Surf. Physicochem. Eng. Asp.*, vol. 302, no. 1 – 3, pp. 150 – 156, Jul. 2007.

[52] O. Kysilka, M. Rybáčková, M. Skalický, M. Kvíčalová, J. Cvačka, and J. Kvíčala, "Fluorous imidazolium room-temperature ionic liquids based on HFPO trimer," *J. Fluor. Chem.*, vol. 130, no. 7, pp. 629 – 639, Jul. 2009.

[53] J. G. Werner, S. S. Johnson, V. Vijay, and U. Wiesner, "Carbon-Sulfur Composites from Cylindrical and Gyroidal Mesoporous Carbons with Tunable Properties in Lithium-Sulfur Batteries," *Chem. Mater.*, vol. 27, no. 9, pp. 3349 – 3357, May 2015.

[54] A. Modaressi, H. Sifaoui, M. Mielcarz, U. Domańska, and M. Rogalski, "Influence of the molecular structure on the aggregation of imidazolium ionic liquids in aqueous solutions," *Colloids Surf. Physicochem.*

Eng. Asp., vol. 302, no. 1–3, pp. 181–185, Jul. 2007.

[55] S. Zhang, Q. Zhang, Y. Zhang, Z. Chen, M. Watanabe, and Y. Deng, "Beyond solvents and electrolytes: Ionic liquids-based advanced functional materials," *Prog. Mater. Sci.*, vol. 77, pp. 80–124, Apr. 2016.

[56] S. Mohamad, H. Surikumaran, M. Raoov, T. Marimuthu, K. Chandrasekaram, and P. Subramaniam, "Conventional Study on Novel Dicationic Ionic Liquid Inclusion with β-Cyclodextrin," *Int. J. Mol. Sci.*, vol. 12, no. 12, pp. 6329–6345, Sep. 2011.

[57] M. Claros, T. A. Graber, I. Brito, J. Albanez, and J. A. GavíN, "SYNTHESIS AND THERMAL PROPERTIES OF TWO NEW DICATIONIC IONIC LIQUIDS," *J. Chil. Chem. Soc.*, vol. 55, no. 3, pp. 396–398, 2010.

[58] W. Shibayama, A. Narita, N. Matsumi, and H. Ohno, "Design and evaluation of imidazolium cation-based ionic liquids having double-armed anions for selective cation conduction," *J. Organomet. Chem.*, vol. 694, no. 11, pp. 1642–1645, May 2009.

[59] R. Fareghi-Alamdari, F. Ghorbani-Zamania, and M. Shekarriz, "Synthesis and Thermal Characterization of Mono and Dicationic Imidazolium Pyridinium Based Ionic Liquids," *Orient. J. Chem.*, vol. 31, no. 2, pp. 1127–1132, Jun. 2015.

[60] J. Wang, S. Zhang, J. Wang, X. Lu, and Q. Zhou, Eds. Berlin, Heidelberg and H. Wang, "Aggregation in Systems of Ionic Liquids, in Structures and Interactions of Ionic Liquids" vol. 151, *Springer Berlin Heidelberg*, 2014, pp. 39–77.

[61] Q. Zhang, S. Liu, Z. Li, J. Li, Z. Chen, R. Wang, L. Lu, and Y. Deng, "Novel Cyclic Sulfonium-Based Ionic Liquids: Synthe-

sis, Characterization, and Physicochemical Properties," *Chem. -Eur. J.*, vol. 15, no. 3, pp. 765–778, Jan. 2009.

[62] Z. Zhang, H. Zhou, L. Yang, K. Tachibana, K. Kamijima, and J. Xu, "Asymmetrical dicationic ionic liquids based on both imidazolium and aliphatic ammonium as potential electrolyte additives applied to lithium secondary batteries," *Electrochimica Acta*, vol. 53, no. 14, pp. 4833–4838, May 2008.

[63] Y. -X. Yin, S. Xin, Y. -G. Guo, and L. - J. Wan, "Lithium-Sulfur Batteries: Electrochemistry, Materials, and Prospects," *Angew. Chem. Int. Ed.*, vol. 52, no. 50, pp. 13186–13200, Dec. 2013.

[64] J. Lu, F. Yan, and J. Texter, "Advanced applications of ionic liquids in polymer science," *Prog. Polym. Sci.*, vol. 34, no. 5, pp. 431–448, May 2009.

[65] J. L. Anderson, R. Ding, A. Ellern, and D. W. Armstrong, "Structure and Properties of High Stability Geminal Dicationic Ionic Liquids," *J. Am. Chem. Soc.*, vol. 127, no. 2, pp. 593–604, Jan. 2005.

[66] P. Kubisa and T. Biedroń, "Poly (oxyethylene) s terminated at both ends with phosphonium ion end groups, 2. Properties," *Macromol. Chem. Phys.*, vol. 197, no. 1, pp. 31–40, Jan. 1996.

[67] Y. Nakai, "Ion conduction in molten salts prepared by terminal-charged PEO derivatives," *Solid State Ion.*, vol. 113–115, no. 1–2, pp. 199–204, Dec. 1998.

[68] Y. Chen, K. Zhuo, J. Chen, and G. Bai, "Volumetric and viscosity properties of dicationic ionic liquids in (glucose + water) solutions at T = 298.15K," *J. Chem. Thermodyn.*, vol. 86, pp. 13–19, Jul.

2015.

[69] S. Li, P. Zhang, F. Fulvio Pasquale, C. Hillesheim Patrick, G. Feng, S. Dai, and T. Cummings Peter, "Enhanced performance of dicationic ionic liquid electrolytes by organic solvents," *J. Phys. Condens. Matter*, vol. 26, no. 28, p. 284105, Jul. 2014.

[70] H. Yang, J. Liu, Y. Lin, J. Zhang, and X. Zhou, "PEO-imidazole ionic liquid-based electrolyte and the influence of NMBI on dye-sensitized solar cells," *Electrochimica Acta*, vol. 56, no. 18, pp. 6271–6276, Jul. 2011.

[71] H. Tadesse, A. J. Blake, N. R. Champness, J. E. Warren, P. J. Rizkallah, and P. Licence, "Supramolecular architectures of symmetrical dicationic ionic liquid based systems," *CrystEngComm*, vol. 14, no. 15, p. 4886, 2012.

[72] A. N. Tran, T. -N. Van Do, L. –P. M. Le, and T. N. Le, "Synthesis of new fluorinated imidazolium ionic liquids and their prospective function as the electrolytes for lithium-ion batteries," *J. Fluor. Chem.*, vol. 164, pp. 38–43, Aug. 2014.

[73] M. B. Herath, T. Hickman, S. E. Creager, and D. D. DesMarteau, "A new fluorinated anion for room-temperature ionic liquids," *J. Fluor. Chem.*, vol. 132, no. 1, pp. 52–56, Jan. 2011.

[74] X. Li, D. W. Bruce, and J. M. Shreeve, "Dicationic imidazolium-based ionic liquids and ionic liquid crystals with variously positioned fluoro substituents," *J. Mater. Chem.*, vol. 19, no. 43, p. 8232, 2009.

[75] Q. Cheng, Z. Cui, J. Li, S. Qin, F. Yan, and J. Li, "Preparation and performance of polymer electrolyte based on poly (vinylidene fluoride)/polysulfone blend membrane via thermally induced phase

separation process for lithium ion battery," *J. Power Sources*, vol. 266, pp. 401–413, Nov. 2014.

[76] Y. Zhu, M. D. Casselman, Y. Li, A. Wei, and D. P. Abraham, "Perfluoroalkyl-substituted ethylene carbonates: Novel electrolyte additives for high-voltage lithium-ion batteries," *J. Power Sources*, vol. 246, pp. 184–191, Jan. 2014.

[77] S. Y. Xiao, Y. Q. Yang, M. X. Li, F. X. Wang, Z. Chang, Y. P. Wu, and X. Liu, "A composite membrane based on a biocompatible cellulose as a host of gel polymer electrolyte for lithium ion batteries," *J. Power Sources*, vol. 270, pp. 53–58, Dec. 2014.

[78] M. Yoshizawa and H. Ohno, "Synthesis of molten salt-type polymer brush and effect of brush structure on the ionic conductivity," *Electrochimica Acta*, vol. 46, no. 10–11, pp. 1723–1728, Mar. 2001.

[79] S. B. Brijmohan, S. Swier, R. A. Weiss, and M. T. Shaw, "Synthesis and Characterization of Cross-linked Sulfonated Polystyrene Nanoparticles," *Ind. Eng. Chem. Res.*, vol. 44, no. 21, pp. 8039–8045, Oct. 2005.

[80] B. Mandal, "New lithium salts for rechargeable battery electrolytes," *Solid State Ion.*, vol. 175, no. 1–4, pp. 267–272, Nov. 2004.

[81] M. Ulaganathan, R. Nithya, and S. Rajendr, "Surface Analysis Studies on Polymer Electrolyte Membranes Using Scanning Electron Microscope and Atomic Force Microscope," *Scanning Electron Microscopy*, V. Kazmiruk, Ed. InTech, 2012.

[82] E. Paillard, C. Iojoiu, F. Alloin, J. Guindet, and J. -Y. Sanchez, "Poly (oxyethylene) electrolytes based on lithium pentafluorobenzene sulfonate," *Electrochimica Acta*, vol. 52, no. 11, pp. 3758–

3765, Mar. 2007.

[83] M. S. Kim, S. K. Kim, J. Y. Lee, S. H. Cho, K. -H. Lee, J. Kim, and S. -S. Lee, "Synthesis of polystyrene nanoparticles with monodisperse size distribution and positive surface charge using metal stearates," *Macromol. Res.*, vol. 16, no. 2, pp. 178 – 181, Feb. 2008.

[84] R. Bouchet, S. Maria, R. Meziane, A. Aboulaich, L. Lienafa, J. -P. Bonnet, T. N. T. Phan, D. Bertin, D. Gigmes, D. Devaux, R. Denoyel, and M. Armand, "Single-ion BAB triblock copolymers as highly efficient electrolytes for lithium-metal batteries," *Nat. Mater.*, vol. 12, no. 5, pp. 452 – 457, Mar. 2013.

[85] S. Feng, D. Shi, F. Liu, L. Zheng, J. Nie, W. Feng, X. Huang, M. Armand, and Z. Zhou, "Single lithium-ion conducting polymer electrolytes based on poly [(4-styrenesulfonyl) (trifluoromethanesulfonyl) imide] anions," *Electrochimica Acta*, vol. 93, pp. 254 – 263, Mar. 2013.

[86] B. K. Mandal and R. Filler, "New fluorine-containing plasticized low lattice energy lithium salt for plastic batteries," *J. Fluor. Chem.*, vol. 126, no. 5, pp. 845 – 848, May 2005.

[87] M. M. Silva, S. C. Barros, M. J. Smith, and J. R. MacCallum, "Characterization of solid polymer electrolytes based on poly (trimethylenecarbonate) and lithium tetrafluoroborate," *Electrochimica Acta*, vol. 49, no. 12, pp. 1887 – 1891, May 2004.

[88] U. Olgun and M. Gülfen, "Synthesis of fluorescence poly (phenylenethiazolo [5, 4-d] thiazole) copolymer dye: Spectroscopy, cyclic voltammetry and thermal analysis," *Dyes Pigments*, vol. 102, pp. 189 – 195, Mar. 2014.

[89] R. Bouchet, S. Maria, R. Meziane, A. Aboulaich, L.

Lienafa, J. -P. Bonnet, T. N. T. Phan, D. Bertin, D. Gigmes, D. Devaux, R. Denoyel, and M. Armand, "Single-ion BAB triblock copolymers as highly efficient electrolytes for lithium-metal batteries," *Nat. Mater.*, vol. 12, no. 5, pp. 452–457, Mar. 2013.

[90] S. M. Ebrahim, M. M. A. -E. Latif, A. M. Gad, and M. M. Soliman, "Cyclic voltammetry and impedance studies of electrodeposited polypyrrole nanoparticles doped with 2 – acrylamido – 2 – methyl – 1 – propanesulfonic acid sodium salt," *Thin Solid Films*, vol. 518, no. 15, pp. 4100–4105, May 2010.

[91] P. K. Sharma, G. Gupta, V. V. Singh, B. K. Tripathi, P. Pandey, M. Boopathi, B. Singh, and R. Vijayaraghavan, "Synthesis and characterization of polypyrrole by cyclic voltammetry at different scan rate and its use in electrochemical reduction of the simulant of nerve agents," *Synth. Met.*, vol. 160, no. 23–24, pp. 2631–2637, Dec. 2010.

[92] O. Geiculescu, Y. Xie, R. Rajagopal, S.. Creager, and D.. DesMarteau, "Dilithium bis [(perfluoroalkyl) sulfonyl] diimide salts as electrolytes for rechargeable lithium batteries," *J. Fluor. Chem.*, vol. 125, no. 8, pp. 1179–1185, Aug. 2004.

[93] A. Chakrabarti, R. Filler, and B. K. Mandal, "Synthesis and properties of a new class of fluorine-containing dilithium salts for lithium-ion batteries," *Solid State Ion.*, vol. 180, no. 40, pp. 1640–1645, Jan. 2010.

[94] D. R. MacFarlane, S. A. Forsyth, J. Golding, and G. B. Deacon, "Ionic liquids based on imidazolium, ammonium and pyrrolidinium salts of the dicyanamide anion," *Green Chem.*, vol. 4, no. 5, pp. 444–448, Oct. 2002.

[95] M. -J. Deng, P. -Y. Chen, T. -I. Leong, I. -W.

Sun, J. -K. Chang, and W. -T. Tsai, "Dicyanamide anion based ionic liquids for electrodeposition of metals," *Electrochem. Commun.*, vol. 10, no. 2, pp. 213–216, Feb. 2008.

[96] K. Dokko, N. Tachikawa, K. Yamauchi, M. Tsuchiya, A. Yamazaki, E. Takashima, J. -W. Park, K. Ueno, S. Seki, N. Serizawa, and M. Watanabe, "Solvate Ionic Liquid Electrolyte for Li-S Batteries," *J. Electrochem. Soc.*, vol. 160, no. 8, pp. A1304–A1310, Jun. 2013.

[97] D. W. Seo, M. K. Parvez, S. H. Lee, J. H. Kim, S. R. Kim, Y. D. Lim, and W. G. Kim, "Synthesis of acetyl imidazolium-based electyrolytes and application for dye-sensitized solar cells," *Electrochimica Acta*, vol. 57, pp. 285–289, Dec. 2011.

[98] B. Yang, C. Li, J. Zhou, J. Liu, and Q. Zhang, "Pyrrolidinium-based ionic liquid electrolyte with organic additive and LiTFSI for high-safety lithium-ion batteries," *Electrochimica Acta*, vol. 148, pp. 39–45, Dec. 2014.

[99] A. Hofmann, M. Schulz, S. Indris, R. Heinzmann, and T. Hanemann, "Mixtures of Ionic Liquid and Sulfolane as Electrolytes for Li-Ion Batteries," *Electrochimica Acta*, vol. 147, pp. 704–711, Nov. 2014.

[100] M. Li, B. Yang, Z. Zhang, L. Wang, and Y. Zhang, "Polymer gel electrolytes containing sulfur-based ionic liquids in lithium battery applications at room temperature," *J. Appl. Electrochem.*, vol. 43, no. 5, pp. 515–521, May 2013.

[101] M. Palacio and B. Bhushan, "Molecularly thick dicationic ionic liquid films for nanolubrication," *J. Vac. Sci. Technol. Vac. Surf. Films*, vol. 27, no. 4, p. 986, 2009.

[102] E. E. L. Tanner, R. R. Hawker, H. M. Yau, A. K.

Croft, and J. B. Harper, "Probing the importance of ionic liquid structure: a general ionic liquid effect on an SNAr process," *Org. Biomol. Chem.*, vol. 11, no. 43, p. 7516, 2013.

[103] E. Alcalde, I. Dinarès, A. Ibáñez, and N. Mesquida, "A Simple Halide-to-Anion Exchange Method for Heteroaromatic Salts and Ionic Liquids," *Molecules*, vol. 17, no. 12, pp. 4007–4027, Apr. 2012.

[104] H. Nakagawa, Y. Fujino, S. Kozono, Y. Katayama, T. Nukuda, H. Sakaebe, H. Matsumoto, and K. Tatsumi, "Application of nonflammable electrolyte with room temperature ionic liquids (RTILs) for lithium-ion cells," *J. Power Sources*, vol. 174, no. 2, pp. 1021–1026, Dec. 2007.

[105] T. Payagala, J. Huang, Z. S. Breitbach, P. S. Sharma, and D. W. Armstrong, "Unsymmetrical Dicationic Ionic Liquids: Manipulation of Physicochemical Properties Using Specific Structural Architectures," *Chem. Mater.*, vol. 19, no. 24, pp. 5848–5850, Nov. 2007.

[106] C. P. Fredlake, J. M. Crosthwaite, D. G. Hert, S. N. V. K. Aki, and J. F. Brennecke, "Thermophysical Properties of Imidazolium-Based Ionic Liquids," *J. Chem. Eng. Data*, vol. 49, no. 4, pp. 954–964, Jul. 2004.

[107] J. -A. Choi, E. -G. Shim, B. Scrosati, and D. -W. Kim, "Mixed Electrolytes of Organic Solvents and Ionic Liquid for Rechargeable Lithium-Ion Batteries," *Bull. Korean Chem. Soc.*, vol. 31, no. 11, pp. 3190–3194, Nov. 2010.

[108] B. S. Lalia, N. Yoshimoto, M. Egashira, and M. Morita, "A mixture of triethylphosphate and ethylene carbonate as a safe additive for ionic liquid-based electrolytes of lithium ion batteries," *J. Power Sources*, vol. 195, no. 21, pp. 7426–7431, Nov. 2010.

[109] C. Arbizzani, G. Gabrielli, and M. Mastragostino, "Thermal stability and flammability of electrolytes for lithium-ion batteries," *J. Power Sources*, vol. 196, no. 10, pp. 4801–4805, May 2011.

[110] M. Montanino, M. Moreno, M. Carewska, G. Maresca, E. Simonetti, R. Lo Presti, F. Alessandrini, and G. B. Appetecchi, "Mixed organic compound-ionic liquid electrolytes for lithium battery electrolyte systems," *J. Power Sources*, vol. 269, pp. 608–615, Dec. 2014.

[111] J. S. Lee, N. D. Quan, J. M. Hwang, J. Y. Bae, H. Kim, B. W. Cho, H. S. Kim, and H. Lee, "Ionic liquids containing an ester group as potential electrolytes," *Electrochem. Commun.*, vol. 8, no. 3, pp. 460–464, Mar. 2006.

[112] K. Yin, Z. Zhang, X. Li, L. Yang, K. Tachibana, and S. Hirano, "Polymer electrolytes based on dicationic polymeric ionic liquids: application in lithium metal batteries," *J Mater Chem A*, vol. 3, no. 1, pp. 170–178, 2015.

[113] J. Xiang, F. Wu, R. Chen, L. Li, and H. Yu, "High voltage and safe electrolytes based on ionic liquid and sulfone for lithium-ion batteries," *J. Power Sources*, vol. 233, pp. 115–120, Jul. 2013.

[114] K. Yin, Z. Zhang, L. Yang, and S. -I. Hirano, "An imidazolium-based polymerized ionic liquid via novel synthetic strategy as polymer electrolytes for lithium ion batteries," *J. Power Sources*, vol. 258, pp. 150–154, Jul. 2014.

[115] R. Kido, K. Ueno, K. Iwata, Y. Kitazawa, S. Imaizumi, T. Mandai, K. Dokko, and M. Watanabe, "Li + Ion Transport in Polymer Electrolytes Based on a Glyme-Li Salt Solvate Ionic Liquid," *Electrochimica Acta*, vol. 175, pp. 5–12, Sep. 2015.

[116] V. Baranchugov, E. Markevich, E. Pollak, G. Salitra,

and D. Aurbach, "Amorphous silicon thin films as a high capacity anodes for Li-ion batteries in ionic liquid electrolytes," *Electrochem. Commun.*, vol. 9, no. 4, pp. 796–800, Apr. 2007.

[117] L. X. Yuan, J. K. Feng, X. P. Ai, Y. L. Cao, S. L. Chen, and H. X. Yang, "Improved dischargeability and reversibility of sulfur cathode in a novel ionic liquid electrolyte," *Electrochem. Commun.*, vol. 8, no. 4, pp. 610–614, Apr. 2006.

[118] F. Pagano, C. Gabler, P. Zare, M. Mahrova, N. Dorr, R. Bayon, X. Fernandez, W. Binder, M. Hernaiz, E. Tojo, and A. Igartua, "Dicationic ionic liquids as lubricants," *Proc. Inst. Mech. Eng. Part J J. Eng. Tribol.*, vol. 226, no. 11, pp. 952–964, Nov. 2012.

[119] H. Yoon, G. H. Lane, Y. Shekibi, P. C. Howlett, M. Forsyth, A. S. Best, and D. R. MacFarlane, "Lithium electrochemistry and cycling behaviour of ionic liquids using cyano based anions," *Energy Environ. Sci.*, vol. 6, no. 3, p. 979, 2013.

[120] P. C. Howlett, D. R. MacFarlane, and A. F. Hollenkamp, "High Lithium Metal Cycling Efficiency in a Room-Temperature Ionic Liquid," *Electrochem. Solid-State Lett.*, vol. 7, no. 5, p. A97, 2004.

[121] P. C. Howlett, D. R. MacFarlane, and A. F. Hollenkamp, "High Lithium Metal Cycling Efficiency in a Room-Temperature Ionic Liquid," *Electrochem. Solid-State Lett.*, vol. 7, no. 5, p. A97, 2004.

[122] S. B. Aher and P. R. Bhagat, "Convenient synthesis of imidazolium based dicationic ionic liquids," *Res. Chem. Intermed.*, Jan. 2016.

[123] R. Hagiwara and Y. Ito, "Room temperature ionic liquids of alkylimidazolium cations and fluoroanions," *J. Fluor. Chem.*, vol. 105, no. 2, pp. 221–227, Sep. 2000.